NATURE'S PATTERNS

A Tapestry in Three Parts

PHILIP BALL

Nature's Patterns is a trilogy composed of
Shapes, Flow, and Branches

OXFORD
UNIVERSITY PRESS

OXFORD
UNIVERSITY PRESS

Great Clarendon Street, Oxford OX2 6DP

Oxford University Press is a department of the University of Oxford.
It furthers the University's objective of excellence in research, scholarship,
and education by publishing worldwide in

Oxford New York

Auckland Cape Town Dar es Salaam Hong Kong Karachi
Kuala Lumpur Madrid Melbourne Mexico City Nairobi
New Delhi Shanghai Taipei Toronto

With offices in

Argentina Austria Brazil Chile Czech Republic France Greece
Guatemala Hungary Italy Japan Poland Portugal Singapore
South Korea Switzerland Thailand Turkey Ukraine Vietnam

Oxford is a registered trade mark of Oxford University Press
in the UK and in certain other countries

Published in the United States
by Oxford University Press Inc., New York

© Philip Ball 2009

The moral rights of the author have been asserted
Database right Oxford University Press (maker)

First published 2009
Published in paperback 2011

British Library Cataloguing in Publication Data

Data available

Library of Congress Cataloging in Publication Data

Data available

Typeset by SPI Publisher Services, Pondicherry, India
Printed in Great Britain
on acid-free paper by
Clays Ltd., St Ives plc

ISBN 978-0-19-960488-3

1 3 5 7 9 10 8 6 4 2

Branching

Seeing the splayed, forking channels of rivers, natural philosophers were reminded of veins and arteries. These in turn speak of trees—and why not, for all are networks that distribute vital fluids. What are the rules that make branches? Why does a lightning bolt seek many routes from heaven to earth, or a crack begin to wander and divide? Branching forms find a compromise between disorder and determinism: they hint at a new and peculiar geometry. Yet sometimes order reasserts itself, as when the arms of snowflakes insist on their hexagonality. And if branches reunite, the loops form a web that offers many routes to the same destination. Navigation and dissemination on such networks then depends on the pattern of connections.

Contents

Preface and Acknowledgements

After my 1999 book *The Self-Made Tapestry: Pattern Formation in Nature* went out of print, I'd often be contacted by would-be readers asking where they could get hold of a copy. That was how I discovered that copies were changing hands in the used-book market for considerably more than the original cover price. While that was gratifying in its way, I would far rather see the material accessible to anyone who wanted it. So I approached Latha Menon at Oxford University Press to ask about a reprinting. But Latha had something more substantial in mind, and that is how this new trilogy came into being. Quite rightly, Latha perceived that the original *Tapestry* was neither conceived nor packaged to the best advantage of the material. I hope this format does it more justice.

The suggestion of partitioning the material between three volumes sounded challenging at first, but once I saw how it might be done, I realized that this offered a structure that could bring more thematic organization to the topic. Each volume is self-contained and does not depend on one having read the others, although there is inevitably some cross-referencing. Anyone who has seen *The Self-Made Tapestry* will find some familiar things here, but also plenty that is new. In adding that material, I have benefited from the great generosity of many scientists who have given images, reprints and suggestions. I am particularly grateful to Sean Carroll, Iain Couzin, and Andrea Rinaldo for critical readings of some of the new text. Latha set me more work than I'd perhaps anticipated, but I remain deeply indebted to her for her vision of what these books might become, and her encouragement in making that happen.

Philip Ball

London, October 2007

A Winter's Tale

The Six-Pointed Snowflake

The followers of Pythagoras believed many strange things, among them that one should not eat beans or break bread, should not pluck a garland and should not allow swallows to land on one's roof. They sound like a bunch of crackpot mystics, but in fact Pythagoreanism has, through its influence on Plato, provided a recurrent theme in Western rationalist thought: the idea that the universe is fundamentally geometric, so that all natural phenomena display a harmony based on number and regularity. Pythagoras is said to have discovered the relationship between proportion and musical harmony, reflected in the way that a plucked string divided by simple length ratios produces pleasing musical intervals. The 'music of the spheres'—celestial harmonies generated by the heavenly bodies according to the sizes of their orbits—is ultimately a Pythagorean concept.

'All things are numbers', said Pythagoras, but it is not easy now to comprehend what he meant by this statement. In some fashion, he believed that integers were building blocks from which the world was constructed. Bertrand Russell is probably imposing too modern a perspective when he interprets the phrase as saying that the world is 'built up of molecules composed of atoms arranged in various shapes', even if, for Plato, those atoms themselves were geometric: cubes, tetrahedra, and other regular shapes that, he said, account for the empirical properties of the corresponding classical elements. All the same, it seems fair to suppose that a Pythagorean would have been less surprised than we are to find spontaneous regularity of pattern and form in the world—five-petalled flowers, faceted crystals—because he would have envisaged this orderliness to be engraved in the very fabric of creation.

The ancient Greeks were not alone in thinking this way. Chinese scholars of long ago were as devoted to the study of nature and mathematics as any of their Western counterparts, and by all appearances they were rather more observant. It was not until the European Middle Ages that the Aristotelian tradition of studying specific natural phenomena for their own sake began to permeate the Western world, prompting the thirteenth-century Bavarian proto-scientist Albertus Magnus to record the 'star-shaped form' of snowflakes, which can be seen with the naked eye. But the Chinese anticipated him by more than a millennium. Around 135 BC, the philosopher Han Ying wrote in his treatise *Moral Discourses Illustrating the Han Text of the 'Book of Songs'* that 'Flowers of plants and trees are generally five-pointed, but those of snow, which are called *ying*, are always six-pointed.' It is a casual reference, as though he is mentioning something that everyone already knew.

Chinese poets and writers in the subsequent centuries took this fact for granted. In the sixth century AD, Hsiao T'ung wrote:

> The ruddy clouds float in the four quarters of the caerulean sky
> And the white snowflakes show forth their six-petalled flowers.

By the seventeenth century, Chinese scholars had become more systematic and scientific in their approach. 'Every year at the end of winter and the beginning of spring I used to collect snow crystals myself and carefully examined them', wrote Hsien Tsai-hang in his *Five Assorted Offering Trays* (c.1600). He may have used a magnifying glass for this work, which led him to conclude that 'all were six-pointed'.

It was no surprise to the Chinese sages that snow crystals were six-pointed, because many of them held a view of nature that was every bit as numerological as that of the Pythagoreans. Still today, numerical schemes provide a central ordering principle in Chinese thought, from the Eightfold Way of Daoism to the 'Four Greats' of Mao's personality cult. In a system of 'correspondences' analogous to that of the Western mystical tradition, the elements were deemed to have numbers associated with them, and as the great philosopher Chu Hsi wrote in the twelfth century, 'Six generated from Earth is the perfected number of Water.' Thus, according to the scholar T'ang Chin, 'when water congeals into flowers they must be six-pointed', because 'six is the true number of Water'.

The problem with this scheme is that it stifles further enquiry: given such an 'explanation' (which we now see as little more than a tautology), there is nothing more to be said. A profound mystery is reduced to a commonplace fact. And so, in the words of sinologist Joseph Needham, 'the Chinese, having found the hexagonal symmetry [of snowflakes], were content to accept it as a fact of nature'.

Here, then, is a rejoinder to the accusation that a scientific attitude is prone to blunt our wonderment at the world. In the mystic's teleological universe, order and pattern are only to be expected: they are part of the Grand Design. There is nevertheless value in such an outlook, which can help to bring to our notice the regularities that exist in nature—we may not see them at all if we do not expect them. In fact, mysticism in all its guises can lead us to perceive *too much* order, making us prone to seeing significance where there is only the play of chance. The human mind seems to be predisposed to this error, for pattern recognition is an essential survival tool and it seems we must resign ourselves to living with its tiresome side-effects, from numerology to 'faces' on the surface of Mars.

But although the mystical Platonic vision of a geometric, ordered universe helped prepare the ground for early Western science, it needed to be replaced by something more empirical, more discerning and sceptical, before we could truly begin to understand how the world works. The snowflake offers a delightful illustration of that process. For it is only when we start to regard these ice crystals as things in themselves, and not as symbols of some deeper principle of nature, that we can truly appreciate how astonishing they are. Their elegance and beauty is, I believe, unrivalled in the natural world, and even Bach would have been silenced by the invention with which they play variations on a simple theme, this interplay of 'sixness' and 'branchingness' in which symmetry seems to be taken about as far as it can tolerate (Fig. 1.1 and Plate 1). They are formed from chaos, from the random swirling of water vapour that

FIG. 1.1 The snowflake displays an urge for branching growth played out with exquisite hexagonal symmetry. (Photo: Ken Libbrecht, California Institute of Technology.)

condenses molecule by molecule, with no template to guide them. Whence this branchingness? Wherefore this sixness?

KEPLER'S BALLS

In the mechanistic worldview that emerged in the West during the wane of the Renaissance, an appeal to numerology could not suffice to account for the remarkable symmetry of the snowflake. The spirit of the age insisted on causative forces that dictated how things happened in their own terms. One could concede that God set the forces at play while insisting that, on a day-to-day basis, they were all He had to work with.

Snowflakes interested the Englishman Thomas Hariot, who noted in his private manuscripts in 1591 that they have six points. Hariot was a masterful mathematician, noted for his contributions to algebra, but his enthusiasms showed the characteristic magpie diversity of the Elizabethan intellectual, among them astronomy, astrology, and linguistics. He tutored Walter Raleigh in mathematics, and when Raleigh set out on a voyage to the New World in 1585 he employed Hariot as navigator. Together they sailed to the land that Raleigh named in honour of his Virgin Queen: Virginia. On the voyage, Raleigh sought Hariot's expert advice about the most efficient way to stack cannonballs on deck.

The question led Hariot to the beginnings of a theory about the close-packing of spheres. Some time between 1606 and 1608 he communicated his thoughts to a fellow astronomer, the German Johannes Kepler, who enjoyed the patronage of the Holy Roman Emperor Rudolph II at his illustrious court in Prague. Most of the correspondence between Kepler and Hariot concerns the refraction of light and the origin of rainbows, but they also discussed atomism: what are atoms, and can empty space come between them? This was an ancient theme, prompted by the belief that nature abhors a vacuum, but it seemed then to be as irresolvable as ever. The issue of how atoms sat against one another brought Hariot back to Raleigh's cannonballs, and he asked what Kepler thought about the matter. In 1611 Kepler wrote a short treatise in which he speculated that the familiar cannonball stacking, which disports the balls in a hexagonal, honeycomb array, is the densest arrangement there can be. The hexagonal packing 'will be the tightest possible', he wrote, 'so that in no other arrangement could more pellets be stuffed into the same container'.* The booklet in which this assertion was contained was a New Year's gift from Kepler to his patron Johann Matthäus Wacker von Wackenfels: seasonably so, for its title indicates the object towards which Kepler's thoughts on close-packing became directed. It was called *On the Six-Cornered Snowflake*.

'There must be a cause why snow has the shape of a six-cornered starlet', Kepler says. 'It cannot be chance. Why always six? The cause is not to be looked for in the material, for vapour is formless and flows, but in an agent.' But Kepler does not claim that he can solve the mystery; indeed, his booklet is a rather charming study in bafflement, full of false

---

*Kepler's conjecture remained just that for nearly four centuries. It was proven to be true by the American mathematician Thomas Hales in 1998.

trails and head-scratching. Nonetheless, it contains the seed of an import-ant idea. Prompted by his discussions with Hariot, Kepler began to think about the geometrical shapes that bodies will adopt if their constituent particles are close-packed like cannonballs. He suggested that the hex-agonal symmetry he had seen in snowflakes that he collected and ob-served that very winter might stem from the stacking of 'globules' of water. These globules are not in themselves atoms; rather, he said, 'vapour coagulates into globules of a definite size, as soon as it begins to feel the onset of cold'. They are like little droplets, and, as such, are perfectly spherical.

Yet in the end Kepler rejects this idea, for he notes that balls can be packed into other regular patterns too—notably square arrays—and yet four-pointed snowflakes are never observed. He remarks that flowers commonly display five-pointed heads (a notion I explored in Book I), which he attributes to a 'formative faculty' or plant soul. But 'to imagine an individual soul for each and any starlet of snow is utterly absurd', Kepler wrote, 'and therefore the shapes of snowflakes are by no means to be deduced from the operation of soul in the same way as with plants.' So how does water vapour acquire a formative faculty? It must, in the end, be God's work—which sounds like a capitulation, but in fact reflects the semi-mystical belief common among early seventeenth-century philo-sophers that nature is imbued with 'hidden' forces that shape its forms. Yet what purpose could be served by this symmetrical expression of a gaseous formative faculty? There is none, Kepler decides: 'No purpose can be observed in the shaping of a snowflake . . . [the] formative reason does not act only for a purpose, but also to adorn . . . [it] is in the habit also of playing with the passing moment.' In this seemingly whimsical conclusion we can discern something valid and profound—for, as I hope this trilogy will show, nature does indeed seem to have an intrinsic pattern-forming tendency that it exercises as though from some irrepressible urge. Kepler even hints inadvertently at the way this impulse can act in living organ-isms in apparent defiance of the strict utilitarianism that Darwinism later seemed to dictate.

Despite its inconclusiveness, Kepler's treatise on the snowflake estab-lished the idea that the geometric shapes of crystals are related to the ordered arrangements of their component units. From this elementary

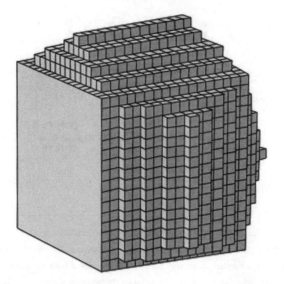

FIG. 1.2 Early crystallographers such as René Just Häuy, from whose book *Traité de Minéralogie* (1801) this illustration comes, explained the faceted shapes of crystals in terms of the packing of their component atoms.

notion came the science of crystallography, beginning in the late eighteenth century, in which the faceted nature of mineral crystals is explained in terms of close-packing of their atoms and molecules (Fig. 1.2). And what is more, his invocation of an almost vitalistic principle behind the growth of snowflakes, redolent of (if not the same as) the 'soul' that guides the growth of plants, captures something of the confusion that snowflakes provoke. The sixness, the hexagonal symmetry, speaks of crystals, of a regularity so perfect that it appears barren. But the branchingness hints at life and growth, at something vegetative and vital.

René Descartes, the arch-mechanist of the early Enlightenment, could not resist the allure of snowflakes. He sketched them in 1637 for his study of meteorology, *Les Météores*, where he recorded rarer varieties alongside the six-pointed stars (Fig. 1.3):

> After this storm cloud, there came another, which produced only little roses or wheels with six rounded semicircular teeth ... which

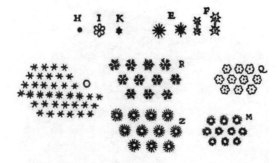

FIG. 1.3 Drawings of snowflakes by René Descartes in 1637.

were quite transparent and quite flat . . . and formed as perfectly and as symmetrically as one could possibly imagine. There followed, after this, a further quantity of such wheels joined two by two by an axle, or rather, since at the beginning these axles were quite thick, one could as well have described them as little crystal columns, decorated at each end with a six-petalled rose a little larger than their base. But after that there fell more delicate ones, and often the roses or stars at their ends were unequal. But then there fell shorter and progressively shorter ones until finally these stars completely joined, and fell as double stars with twelve points or rays, rather long and perfectly symmetrical, in some all equal, in other alternately unequal.

We can recognize in this vivid description some of the unusual forms that have been found in snowflakes, such as prismatic columns with end-caps, like elaborate sundials, and twelve-pointed stars in which two hexagonal flakes have become fused (Fig. 1.4).

The English scientist Robert Hooke had the advantage of a microscope in preparing illustrations of snowflakes for his famous *Micrographia* (1665), where he shows that the 'flowers' are not just six-pointed but branch repeatedly, in a hierarchical manner (Fig. 1.5a). The organic associations of these ice crystals are very apparent in the drawings by the Italian astronomer Giovanni Domenico Cassini in 1692, where they look almost leafy (Fig. 1.5b). The biologist Thomas Huxley acknowledged this aspect in 1869,

FIG. I.4 Twelve-pointed snowflakes are formed when two normal six-pointed varieties fuse together at their centres, rotated in relation to one another by about 30°. (Photo: Ken Libbrecht, California Institute of Technology.)

when he called snowflakes 'frosty imitations of the most complex forms of vegetable foliage'. Huxley's comments appeared in an essay on 'the physical basis of life', in which he strove like a good positivist to quell any notion of a vital force that animated organic matter and made it fundamentally different from the inorganic world. To Huxley, the 'organic' forms of snowflakes provided evidence that the complex shapes of the biological world need not compel the scientist to invoke some mysterious vitalistic sculpting mechanism, since something of that nature surely did not operate in the simple process of the freezing of water:

> We do not assume that a something called 'aquosity' entered into and took possession of the oxide of hydrogen as soon as it was formed, and then guided the aqueous particles to their places in the facets of the crystal, or amongst the leaflets of the hoar-frost.

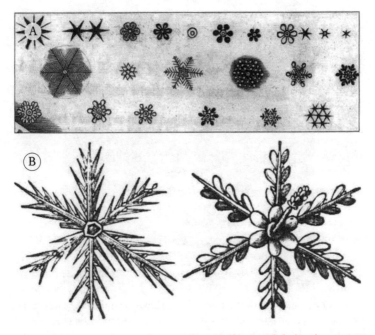

FIG. 1.5 Using an early microscope, Robert Hooke recorded the characteristic 'Christmas-tree' branching patterns of snowflakes (*a*). Giovanni Domenico Cassini's drawings from 1692 seem to make reference to their resemblance to plants (*b*).

It was a reasonable enough assertion, but it surely begs the question: if there is nothing 'organic' about the formation of the snowflake, why then do they look so tantalizingly as though there is?

FLAKES FROZEN ON FILM

As Descartes hinted, the shapes of snowflakes can evolve and mutate as the weather changes. Friedrich Martens, on board a ship travelling from Spitzbergen in Norway to Greenland in 1675, noticed that different meteorological conditions produce different kinds of flake. It takes an Arctic chill to condense the best, most symmetrical snowflakes, as the English explorer William Scoresby noted in his *Account of the Arctic Regions with a History and Description of the Northern Whale-Fishery* in 1820. Scoresby took

FIG. 1.6   In 1820, explorer William Scoresby made accurate drawings of the snowflakes he observed during a trip to the Arctic.

the observations of snowflakes to a new standard of detail and accuracy, recording a wide range of different shapes (Fig. 1.6). One of the most charming of nineteenth-century records was that produced in 1864 by a minister's wife in Maine named Frances Knowlton Chickering, who used, if not invented, the trick now popular at Christmas of cutting out doily-style snowflakes from folded paper. Chickering's paper flakes were masterpieces of dexterity: from memory of her first-hand observations, she clipped out delicate frond-like branches and pasted the results into her *Cloud Crystals: A Snow-Flake Album*, which implicitly acknowledged the 'artistry' of natural phenomena that the biologist Ernst Haeckel was later to celebrate in his drawings of marine life, as we saw in Book I.

The accuracy of all these visual records of snowflakes was limited not only by the power of the magnifying glass or microscope but also by the artist's inevitable tendency to simplify, idealize, and interpolate these complex geometric forms. That problem was avoided once researchers found a way to marry the new art of photography to the power of the microscope. Microphotography was already a well-established technique by the late nineteenth century, and one of its most inventive practitioners was a Vermont farmer named Wilson Bentley. Between 1885 and 1931, Bentley captured over 5,000 images of snowflakes on photographic plates, constituting one of the most comprehensive surveys of their astonishing variety and beauty (Fig. 1.7). In the late 1920s Bentley compiled 2,000 of his photographs into a book entitled *Snow Crystals* in collaboration with

FIG. 1.7 The collection of snowflake photographs amassed by Wilson Bentley in the four decades after 1885 still stands as the most remarkable record of their endlessly varied forms.

William J. Humphreys, a physicist working for the US Weather Bureau. Bentley died only a few weeks after the book was published in November 1931, allegedly after contracting pneumonia during one of his forays into the New England winter.

Snow Crystals is rightly regarded as a work of wonder, but it is more than that. The scientist, gazing at page after page of seemingly infinite variety on the theme of the six-pointed flower of ice, faces a mystery of an order not previously encountered in the non-living world. Not only were the forms indescribably complex, but there was no end to them.

Bentley's album was pure description, to which Humphreys could add rather little in the way of hard science. But in the 1930s the book inspired a Japanese nuclear physicist named Ukichiro Nakaya, working at the University of Hokkaido, to consider the question of snowflake growth in a rather more analytical spirit. He made the first systematic attempt to discover the factors that influenced snowflake growth, leading to the many different *families* of shapes that had been seen by Scoresby and others in the natural environment. Nakaya realized that snowflakes fall into several distinct categories, and he constructed a laboratory for exploring the conditions that generated these different classes of shape.

It was uncomfortable work: Nakaya's wooden-walled lab could be cooled to $-30\,°C$, and he worked in padded clothing with a mask to protect his face. Snowflakes grow slowly as they fall through the atmosphere, but Nakaya could not recreate this long descent in the lab, so instead he decided to reverse the situation: to hold the snowflake fixed and to let cold, moist air pass over it in a steady stream. The question was, how do you hold onto a snowflake? Nakaya experimented with many different kinds of filament for immobilizing a growing crystal of ice, but most of them simply became coated with frost. He finally found that the experiment worked best with a strand of rabbit hair, on which the natural oils suppressed the simultaneous nucleation of many ice crystals at once (Fig. 1.8). Using this equipment, Nakaya and his co-workers found that the shapes of the individual crystals changed as two key factors were altered: the temperature and humidity of the air. At low humidity, the crystals did not develop the six frond-like arms of classic snowflakes, but took on more compact forms: hexagonal plates and prisms. These shapes persisted even in moister air if it was very cold (below about $-20\,°C$). At higher

FIG. 1.8   Snowflakes made artificially by Ukichiro Nakaya in the 1930s. In the image on the left, the rabbit's hair on which the crystals are nucleated is still visible.

temperatures, however, increasing the humidity tended to increase the delicacy and complexity of the snowflakes, giving rise to the highly branched star forms. In a temperature range between about −3 and −5 °C, needle-like crystals appeared instead (Fig. 1.9).

Nakaya collected his findings in an album of images clearly indebted to Bentley and Humphreys, called *Snow Crystals: Natural and Artificial* (1954). His studies brought some order to the ice menagerie, but they did not really bring us any closer to understanding the fundamental mechanism by which a simple process of crystallization, which typically generates a compact prismatic or polyhedral shape, in this case gives us structures that seem to have a life of their own.

As I explained in the previous volumes, the first person to tackle this sort of question about the genesis of complex form within a modern scientific framework was the Scottish zoologist D'Arcy Wentworth Thompson, whose 1917 book *On Growth and Form* set the scene for everything I discuss in this series. Thompson included drawings based on Bentley's photographs in the 1942 revised edition of his book. 'The snow crystal', he wrote, 'is a regular hexagonal plate or thin prism.' But 'ringing her changes on this fundamental form, Nature superadds to the primary hexagon endless combinations of similar plates or prisms, all with identical angles but varying lengths of side; and she repeats, with an exquisite symmetry, about all three axes of the hexagon, whatsoever she may have done for the adornment and elaboration of one.' In other words, all the arms appear to

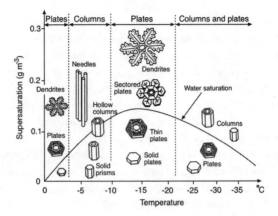

FIG. 1.9 The 'morphology diagram' of snowflakes, showing how their shape changes for different conditions of temperature and humidity (supersaturation).

be identical. 'The beauty of a snow-crystal depends on its mathematical regularity and symmetry', Thompson observed,

> but somehow the association of many variants of a single type, all related but no two the same, vastly increases our pleasure and admiration. Such is the peculiar beauty which a Japanese artist sees in a bed of rushes or a clump of bamboos, especially when the wind's ablowing; and such is the phase-beauty of a flowering spray when it shews every gradation from opening bud to fading flower.

Here it is again: flowers and ice. But even Thompson, like Kepler, could say no more. With all his ideas about forces and equilibria and geometry, he, too, was forced to take recourse in metaphors from the organic world.

ENDLESS BRANCHES

By the time Nakaya's book appeared, scientists had found a way to attack the problem. Although ice seems to be unique in forming highly symmetrical, isolated flakes, many other substances may crystallize as needle-like protrusions punctuated by regular branches, like a single snowflake arm. These structures, known as dendrites (from the Greek for 'tree') are found when molten metals freeze (Fig. 1.10a), when salts precipitate out of a solution, and when metal deposits form on electrically charged electrodes, a process known as electrodeposition and related to electroplating (Fig. 1.10b) (see page 31). Dendrites typically have a rounded tip, like the prow of a boat, behind which side-arms sprout and grow in a Christmas-tree pattern. In general, they appear when the solidification process happens rapidly, as for example when a molten metal is quenched ('undercooled') by being plunged into cold surroundings. That's an important clue. We observed in Book I that complex pattern and form is often generated in processes that take place significantly out of thermodynamic equilibrium—which is to say, when the system is highly unstable. A system in equilibrium does not change; a system out of equilibrium 'seeks' to attain such a stable state if left alone to do so, but can be driven away from this goal by a constant influx of energy. We saw in Book II that convection (the flow of a fluid when heated from below) produces such a non-equilibrium state. A liquid that is abruptly cooled far below its freezing point is another non-equilibrium system, being unstable relative to the solid form of the material. That instability makes change happen rapidly, under which conditions pattern is apt to appear. In contrast, crystals that are formed close

FIG. 1.10 Dendrites formed by rapid solidification of a molten metal (a) and in the electrodeposition of a metal (b). (Photos: a, Lynn Boatner, Oak Ridge National Laboratory, Tennessee. b, Eshel Ben-Jacob, Tel Aviv University.)

to equilibrium—very close to their freezing point, say—grow slowly, and tend instead to develop the familiar compact, faceted shapes.

In 1947 the Russian mathematician G. P. Ivantsov showed theoretically that a metal solidifying rapidly from its molten form may develop needle-like fingers. Ivantsov calculated that the needles have a shape mathematicians called parabolic, with gently curving sides that converge on a blunt tip. This is the same shape as the trajectory of a stone thrown through the air and falling under gravity. Ivantsov showed that in fact all possible types of parabolic needles may be formed, but that the thinner they are, the more rapidly they grow; so thin, needle-like tips should shoot rapidly through the molten metal, while fatter bulges make their way forward at a more ponderous pace.

But in the mid-1970s, Martin Glicksman and co-workers at the Rensselaer Polytechnic Institute in New York performed careful experiments which showed that, instead of a family of parabolic tips, only one single tip shape was seen during rapid solidification of metals. For a fixed degree of undercooling, a particular tip is privileged over the others. For some reason, one of Ivantsov's family of parabolas seems to be special.

The puzzle was even more profound, however, because in 1963 two Americans, William Mullins and Robert Sekerka at Carnegie Mellon University in Pittsburgh, argued that *none* of Ivantsov's parabolas should be stable. They calculated that the slightest disturbance to the growth of a parabolic tip will be self-amplifying, so that small bulges that form by chance on the edge of the crystal grow rapidly into thin fingers. This so-called Mullins–Sekerka instability should cause the tip to sprout a jumble of random branches.

The instability is an example of a positive feedback process—again, we have encountered such things already in Books I and II. It works like this. When a liquid freezes, it releases heat. This is called latent heat, and it is the key to the difference between a liquid and its frozen, solid form at the same temperature. Ice and water can both exist at zero degrees centigrade, but the water can become ice only after it has becomes less 'excited'—its molecules cease their vigorous jiggling motions—by giving up latent heat.

So, in order to freeze, an undercooled liquid has to unload its latent heat. The rate of freezing depends on how quickly heat can be conducted away from the advancing edge of the solid. This in turn depends on how steeply the temperature drops from that in the liquid close to the solidification front to that in the liquid further away: the steeper the gradient in temperature, the faster heat flows down it. (It may seem odd that the liquid close to the freezing front is actually warmer than that further away, but this is simply because the front is where the latent heat is released. Remember that in these experiments all of the liquid has been rapidly cooled below its freezing point but has not yet had a chance to freeze.)

If a bulge develops by chance—because of the random motions of the atoms and molecules, say—on an otherwise flat solidification front, the temperature gradient becomes steeper around the bulge than elsewhere, because the temperature drops over a shorter distance (Fig. 1.11). So latent heat is shed around the bulge more rapidly than it is to either side, and the bulge grows, its apex fastest. This in turn sharpens the tip and speeds its advance even more.

In principle, this instability will amplify any irregularity on the solid front into a growing finger, no matter how small it is. But there is another factor that sets a minimum limit to the width of the fingers. The interface between the solid and the liquid has a surface tension, just like that at the surface of water in a glass. As I explained in Book I, the existence of surface tension means that an interface costs energy: the bigger the surface area, the higher the energetic cost. Surface tension thus encourages surfaces to keep their area as small as possible, and here it tends to 'pull' the solidification front flat. Thanks to this smoothing effect, surface tension

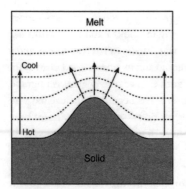

FIG. 1.11 The Mullins–Sekerka instability makes protrusions at the surface of a solidifying material unstable. Because the temperature gradient (shown here as dashed contours of equal temperature) is steeper at the tip of the protrusion, heat is conducted away faster and so solidification proceeds more rapidly here.

suppresses bulges smaller than a certain limit. This means that the Mullins–Sekerka instability produces a characteristic branch-tip width, set by the point at which the narrowing of tips caused by positive feedback is counterbalanced by their cost in surface energy. In other words, the front develops fingers with a certain *wavelength*—a regular pattern with a particular size scale to it, determined by a balance of opposing factors.

In 1977 James Langer at the University of California at Santa Barbara and Hans Müller-Krumbhaar in Jülich, Germany, suggested that the Mullins–Sekerka instability might explain Glicksman's observation of a single parabolic tip being selected from all of those allowed by Ivantsov's theory. The instability will make fat fingers break up into a mass of smaller ones, they said, while surface tension sets a limit on how small and narrow they can become. Perhaps, then, there is an optimal tip width at which these two effects balance, favouring a single 'marginally stable' parabolic tip.

But in the early 1980s, Langer and his co-workers showed that surface tension in fact destroys this neat picture. Its influence makes the tip of the dendrite become cooler than the regions to either side. So the tip starts to slow down, and eventually it forks into two new fingers. These also split subsequently, and so on. This repeated tip-splitting results not in a dendritic growth shape at all, but instead a dense mass of repeatedly forking branches, a pattern known as the dense-branching morphology.

This turned out to be a persistent problem with theories of dendrite growth: they seemed prone to instabilities that led to randomly branched fingering patterns, not the orderly, Christmas-tree shapes of snowflake arms. What these theories were neglecting, in the mistaken belief that it was a mere detail, was the most striking aspect of a snowflake's shape: its hexagonal symmetry. It had been suspected ever since Kepler that this was an echo of the underlying symmetry in the arrangement of constituent particles. But no one had guessed that it was to this symmetry that the dendrites owed their very existence.

THE JOY OF SIX

Why hexagons? Remarkably, the Chinese sages were right: there *is* a sense in which six is the number of water. In 1922 the English physicist William Bragg used his new technique of X-ray crystallography to deduce how water molecules are arranged in an ice crystal. X-rays bouncing off crystals produce a pattern of bright spots that encode the positions of the atoms; Bragg deduced how to calculate backwards from the X-ray pattern to the atomic structure. In this way, he found that the water molecules in ice are linked by weak chemical bonds into hexagonal rings, one molecule at each corner (Fig. 1.12). Thus the crystal structure is dictated not by the shapes of the water molecules themselves but by the way in which they are joined together.

It might seem unlikely that this is the origin of the six-pointed snowflake, since water molecules are *very* much smaller than a snowflake—how could this sixness become so amplified? But as Kepler and the early crystallographers realized, geometric packing of a crystal's constituent units dictates the geometry of the much larger bodies that result. In

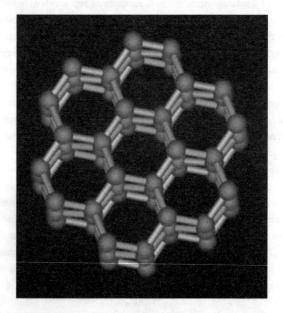

FIG. 1.12 Ice has a structure with hexagonal symmetry at the molecular scale. The water (H_2O) molecules are linked together by weak chemical bonds called hydrogen bonds. In this image the spheres denote the oxygen atoms at the centres of the molecules, and the rods denote the hydrogen bonds that bind molecules together.

essence, water's crystal structure imposes an innate hexagonality on the way the ice crystal grows. Or, as D'Arcy Thompson wrote with his customary elegance, 'these snow-crystals seem to give visible proof of the space-lattice on which their structure is framed'.

The presence of the hexagonal 'space-lattice' means that not all directions are the same for the growing crystal. That is why faceted crystals have the characteristic shapes that they do: the flat facets are simply planes of stacked atoms or molecules, but the reason why certain planes and not others define the crystal's form is that some facets grow faster. This non-equivalence of directions is called *anisotropy*; an isotropic substance is one that looks the same, and behaves in the same way, in all directions.

The anisotropy of crystals means that properties like surface tension differ in different directions. In 1984 Langer and his co-workers showed that, for Ivantsov parabolas growing in certain 'favoured' directions picked out by the anisotropy of the material's crystal structure, surface tension no longer induces a tip-splitting instability—the parabolic tip remains stable as it grows. Thus dendritic branches will grow outwards from an initial crystal 'seed' only in these preferred directions: the snow-flake grows six arms. This special role of anisotropy in stabilizing the growth of a particular needle crystal was identified independently at the same time by David Kessler, Joel Koplik, and Herbert Levine at the University of California at San Diego.

Anisotropy also explains why a dendrite develops side branches. When, by chance, the parabolic tip develops small bulges on its flanks, these may be amplified by the Mullins–Sekerka instability. But again, only bulges that grow in certain directions will be stable. And there is only one kind of dendrite tip, for a given set of growth conditions, which grows fast enough to avoid being overwhelmed by these side branches. So a particular dendrite, with side branches sprouting in particular directions, is uniquely selected from amongst the possible growth shapes.

HOW THE RIGHT ARM KNOWS WHAT
THE LEFT ARM IS DOING

The mind-boggling variety of snowflake forms is therefore the outcome of a tension between chance and necessity. The mechanics of the growth process ensures that the arms will sprout in directions that point to the corners of a hexagon. For any given snowflake, these arms will all grow at the same rate (because, at such a small scale, they all experience the same conditions of temperature and humidity), and so they will have the same length. The side-branches of this six-pointed star are to a degree at the mercy of fate: they may be triggered by the random appearance of tiny irregularities or bulges along the parent arm. Yet they too will always surge outwards in a 'hexagonal' direction. Changes in the prevailing conditions that an individual snowflake experiences as it drifts and falls in the air may trigger simultaneous changes in the growth of all the branches, accounting for how, for example, needle-like arms might develop hexagonal plate-like formations at their tips (Fig. 1.13).

But that doesn't quite explain it all. If pure chance dictates the side-branching of the six arms, why are some snowflakes so amazingly symmetrical even in their fine decorations (Fig. 1.14)? There appears to be something almost magical at play here—each arm seems somehow to know what all the others are doing. Nakaya confessed to being perplexed

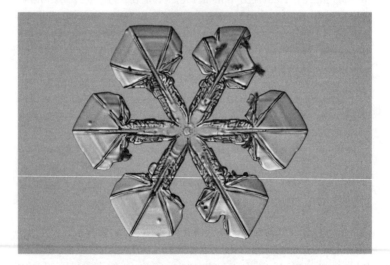

FIG. 1.13   A change in ambient conditions as a snowflake grows in the atmosphere can result in a change in morphology of the branches. Here needle-like branches have developed plate-like tips. (Photo: Ken Libbrecht.)

FIG. 1.14 Why some snowflakes have arms that are essentially identical in all their intricate detail is still somewhat of a mystery. (Photo: Ken Libbrecht, California Institute of Technology.)

by this apparent 'communication' between the branches. 'There is apparently no reason', he said,

> why a similar twig must grow, in the course of the growth of the crystal, from one main branch when a corresponding twig happens to extend from another main branch . . . In order to explain this phenomenon we must suppose the existence of some means which informs other branches of the occurrence of a twig on a point of one branch.

Yet in fact many flakes *do not* have this perfect symmetry: the six arms look roughly identical in their general features, but close inspection reveals differences of detail (Fig. 1.15a). Snowflake expert Ken Libbrecht of the California Institute of Technology, who has taken up the mantle of

FIG. 1.15   The six branches of some snowflakes can be quite different from one another in their fine details, even though at a glance they look symmetrical (*a*). Such flakes can be grown in a computer model of aggregating particles that assigns at random where a new particle gets attached, subject to the constraint that the particle positions must lie on a hexagonal grid (*b*). There is nothing in the rules of the model to ensure that all branches are the same, and indeed they are *not* the same; but our eyes are fooled into seeing more symmetry than there really is by the uniformity of the branching angles. (Photo and image: *a*, Ken Libbrecht, California Institute of Technology; *b*, Gene Stanley, Boston University.)

Nakaya and Bentley and Humpreys in cataloguing the richness and beauty of these crystals by microphotography (his images adorn these pages) says that he commonly rejects thousands of flakes for every one he considers beautiful (which is to say, symmetrical) enough to record. Physicists Johann Nittmann and Gene Stanley propose that almost perfectly regular snowflakes are in fact the exception rather than the rule. They say that the apparent perfection is often illusory: we are fooled into perceiving it simply because each arm has side branches diverging at the same angle and because the 'envelope' of each arm has the same shape. Nittmann and Stanley showed that a model in which particles drift about at random but stick when they make contact* will produce convincing snowflake-like

*We will encounter this model in much more detail in the next chapter, where we will see that it can give rise to a much more widespread and less orderly branching pattern.

shapes if an underlying sixfold symmetry is imposed by constraining the particles to move from site to site on a hexagonal grid, as though hopping between the cells of a honeycomb (Fig. 1.15b). None of the branches in this flake is identical, even though at a glance they look similar. But the general Christmas-tree shape is preserved in all of them, and their lengths are more or less the same, simply because both the main branches and their respective side branches grow at roughly the same rates. The randomness actually ensures this, because it gives no one branch any opportunity to grow faster than the others.

Mathematicians Janko Gravner and David Griffeath found something similar in a more sophisticated model of snowflake growth. Like Nittmann and Stanley's model, this assumes that the crystals grow by accumulating units (much larger than individual water molecules) at the edges from the surrounding vapour, which are constrained to pack together in a flat hexagonal array. But the attachment rules are rather complicated, invoking several parameters whose values can be altered to explore different accumulation conditions. In effect, these rules 'build in' the physical processes involved in crystal growth in a somewhat ad hoc, albeit fairly realistic, way rather than simply letting them emerge (or not) from much simpler rules. Gravner and Griffeath were able to grow a huge variety of snowflakes in their scheme (Fig. 1.16a). In the basic model all the

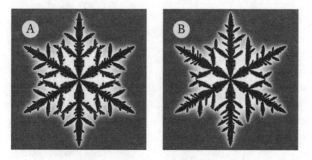

FIG. 1.16 Snowflakes grown in a model developed by Janko Gravner and David Griffeath are remarkably realistic (a). Here the branches are forced to be identical, but when that condition is relaxed by an injection of randomness, the branches still look rather similar (b). (Images: Janko Gravner and David Griffeath, from Gravner and Griffith, 2008.)

arms were forced to be identical, but the researchers could also subvert this by introducing some randomness in the density of the vapour surrounding the crystal. In that case, the snowflakes could still have six branches that looked rather similar at a glance, although differing in detail (Fig. 1.16b). Gravner and Griffeath think that this might be because there are in fact only a relatively small number of stable side-branching designs: the choices the flake has are by no means limitless.

They have extended their model so that it can make three-dimensional crystals, in which case they find not only flakes of startling realism (Fig. 1.17) but also the needle, prism and columnar forms catalogued by Nakaya. While questions of detail remain, it now seems that we are well on the way to decoding the mysteries of 'snow flowers'.

Snowflakes have provided scientists and natural philosophers with one of the clearest indications that complexity of form is not a special

FIG. 1.17 A selection of snowflakes grown in the three-dimensional version of Gravner and Griffeath's model. These display the kinds of decorations, such as ribs and ridges, seen in real flakes. (Images: Janko Gravner and David Griffeath, from Gravner and Griffeath 2009.)

attribute of the living world. Within the snowflake they found an echo of the shapes of trees and flowers, ferns, and starfish: patterns that hint at an exquisite conciliation between geometric purity and organic exuberance. These forms have forced us to re-examine notions of how beauty arises in nature, and how universal processes connect the living and non-living world. 'How full of creative genius is the air in which these are generated!' exclaimed Henry David Thoreau in 1856. 'I should hardly admire more if real stars fell and lodged on my coat.'

TENUOUS MONSTERS

Shapes between Dimensions

I n Mike Leigh's 1976 film *Nuts in May*, a pair of gauche campers named Keith and Candice Marie discover how a kind of modern-day vitalism colours our preconceptions about complex growth and form. They take a trip to a local quarry to look for fossils in the ancient limestone of Dorset in southern England. A quarryman shows the unsuspecting couple a delicate, plant-like pattern traced out in the stone. 'Is that a fossil?', asks Candice Marie, awestruck. 'Ar, most people think that', the Dorset quarryman tells them. 'It's just a mineral.'

It's easy to see why Keith and Candice Marie jumped to the wrong conclusion. The structures they saw are called mineral dendrites (Fig. 2.1), and they look for all the world like the forms we associate with plants— which is of course precisely why they are named after them.* But these filigrees contain no fossil material; they are made of iron or manganese oxides, chemical deposits precipitated when a solution rich in these metal salts was squeezed through cracks in the rocks in the geological past.

---

*These are not the same 'dendrites' as those described in rapidly freezing metals in the previous chapter, which have a more regular, snowflake-arm appearance. Neither, for that matter, have they anything to do with what neuroscientists would recognize by the term, namely the branched extremities of nerve cells in our brain. Metallurgists, geologists and biologists have all been eager to seize upon the tree metaphor, the Greek *dendron*, without much regard for how the word has been applied elsewhere.

FIG. 2.1 A mineral dendrite of manganese oxide found in Bavarian limestone. (Photo: Tamás Vicsek, Eötvös Loránd University, Budapest.)

Mineral dendrites scream 'life' at us because their branched patterns are echoed everywhere in the organic world, from corals to leaf veins to the bronchial structure of the lung. But such branched formations are not by any means unique to biology: think of river networks, lightning, cracks in the ceiling. We will encounter each of these in later chapters. And of course not all branching forms are alike: a snowflake little resembles a naked oak, and I don't think we would mistake this mineral dendrite for a fracture pattern. Botanists and foresters can identify the species of a tree in winter simply from its silhouette on a hilltop. How do they do that—what distinguishes the boughs of an elm from those of a sycamore? I doubt that many botanists could tell you with any great precision. They might mutter something about angles of divergence in the branches, but in the end they just seem able to 'sense' the characteristics of the pattern. Might we, though, hope to establish a taxonomy of branching, and perhaps to find within it a boundary that separates the living from the inorganic?

ORGANIC ROCKS

Keith and Candice Marie were in venerable company. In the early 1670s, Isaac Newton wrote a short tract that is little known now but was probably considered by its author to be the precursor to a (never-written) magnum opus comparable to his *Principia*. It was called *Of Natures obvious lawes and processes in vegetation*, and it was an attempt to sketch out the framework for a 'theory of everything'. The work begins with a description of mineral dendrites, produced artificially—which is to say, alchemically—in Newton's laboratory.

Newton doesn't explain exactly how he made these crystalline growths, but it seems likely that he followed a prescription similar to that of the seventeenth-century German chemist Johann Rudolph Glauber. Glauber described how iron dissolved in 'spirit of salt' (hydrochloric acid) will generate branched deposits of its red oxide when added to a tall flask of 'oil of sand' or 'oil of glass' (potassium silicate). The dendritic structures that can be made this way from precipitating metal salts are now known as silica gardens.

Newton never used that term, but he might be expected to have found it most congenial—for he believed there was something 'vegetative' about metals which led them to grow within the earth as veins that are entirely analogous to trees. The fact that 'metal trees' could be made in the lab apparently confirmed this view.

There was nothing idiosyncratic about the idea, which was shared by many scientists in Newton's day. That metals and salts were formed by vegetative growth was proposed in the early sixteenth century by the Swiss alchemist Paracelsus, who stated that, just as trees have their roots in the soil and grow upwards into the air, so mineral veins have their roots in subterranean water and grow upwards into the earth. Mineral veins, wrote Paracelsus's contemporary Biringuccio, an Italian authority on mining and metallurgy, 'show themselves almost like the veins of blood in the bodies of animals, or the branches of trees spread out in different directions'.

FIG. 2.2 Mineral dendrites drawn by Jean-Jacques Scheuchzer, in his book *Herbarium diluvianum* (1709).

Taking his lead from Paracelsus, the French ceramicist and natural philosopher Bernard Palissy suggested around 1580 that minerals grow from salty 'seeds' that germinate in water. But the idea arose in the late sixteenth century that mineral dendrites might indeed be fossil plants, perhaps produced by inundation in the biblical flood. This position was challenged by the Swiss geologist Jean-Jacques Scheuchzer, whose 1709 book *Herbarium diluvianum* (*Herbarium of the Flood*) contains remarkably accurate drawings of the branched forms of mineral dendrites (Fig. 2.2). Scheuchzer was actually an enthusiastic Diluvianist himself, and he accepted the idea that mineral dendrites were products of the Deluge— but not, he said, as fossil plants. Instead, he suggested a mechanism of formation that, as we shall see, was astonishingly prescient. He explained that if a liquid is squeezed between two flat plates that are then pulled quickly apart, the liquid film is drawn out into a dense array of branches, just like those seen in mineral dendrites, as air pushes its way in. Thus, Scheuchzer explained to the Académie des Sciences in Paris, 'Each time that one finds small trees in stones that can be split easily, which seem to be painted artificially and whose small branches are separated one from the other and never intersect, then one must attribute this arborescence to the injection of a fluid.' But how was this connected to the Flood? Scheuchzer proposed that the seas sloshed violently onto all the lands of the Earth when, in an instant, God stopped the world turning on its axis. The sudden influx of water into the cracks and pores of land-borne rock could have imprinted these delicate mineral fronds.

31

Flood or no flood, Scheuchzer's experiment undermined any need to invoke vegetation as a formative principle for mineral dendrites. By the mid-eighteenth century, the Swedish mineralogist Johann Gottschalk Wallerius was able to write in his influential treatise on minerals that

> There are naturalists who maintain that minerals have a life like that which vegetation enjoys; but since no one has yet been able to see, even with the best microscopes, that these substances have a juice contained in their fibres or veins, since no one has established this view with some evidence, and moreover since it is impossible in general to sustain any notion of life without a circulating juice, one cannot see any basis for ascribing life to minerals, unless one does not want to call living everything that has the faculty to grow and to increase itself.

Already, life was deemed dispensable in making 'organic' forms.

DEPOSIT AT YOUR NEAREST BRANCH

Today it is a rather simple matter to grow ramified crystals such as mineral dendrites in the laboratory. One way is to use the process of electrodeposition that we encountered in the previous chapter. A negatively charged electrode is immersed in a solution of a metal salt, and the pure metal grows at the electrode surface. Metal ions have a positive electrical charge, since they are metal atoms that have lost one or more negatively charged electrons, so they are electrically attracted to the electrode, where they can pick up their missing electrons and revert back to neutral metal atoms. The atoms stack together in a crystal that grows outwards from the electrode surface.

In electroplating, this process is conducted at a low electrode voltage, and the metal film grows slowly, coating the electrode with a smooth, even veneer. But at higher voltages the deposit grows more quickly—that is to say, out of equilibrium. In that case the metal deposit becomes irregular and branches blossom from it (Fig. 2.3a). This no longer looks like a crystalline entity, although close inspection with a microscope reveals that the branches are after all composed of tiny crystals fused together in jumbled disarray (Fig. 2.3b).

FIG. 2.3 A branching metal formation produced by electrochemical deposition onto a central electrode (a). Seen close up under a microscope (b), the branches consist of conglomerates of tiny crystallites oriented at random. (Photos: a, Mitsugu Matsushita, Chuo University; b, Vincent Fleury, Laboratory for Condensed Matter Physics, Palaiseau.)

In 1984 Robert Brady and Robin Ball from the University of Cambridge showed that a theoretical model developed three years earlier by two American physicists, Tom Witten at the Collège de France in Paris and Len Sander at the University of Michigan, could account for the shape of these branched electrodeposits. Witten and Sander hadn't been thinking

33

about electrodeposition at all, however. Their model was supposed to describe something quite different: the way particles of dust form clumps or aggregates as they drift about and stick together in air. Witten and Sander supposed that each particle wanders at random (a process called diffusion) until it hits another particle, whereupon they stick together the moment they touch. This means that particles are added randomly to a growing cluster from all directions. The rate at which the cluster grows depends on how quickly the particles diffuse—how long it takes them to reach the cluster's periphery. For this reason, Witten and Sander called their model *diffusion-limited aggregation* (DLA).

This sounds rather like the way a snowflake grows in the frigid atmosphere: water molecules diffuse until they collide with a growing flake, to which they stick. But there's a crucial difference. In snowflakes the water molecules in the ice crystal stack together in an orderly manner, forming hexagonal rings. This implies that, once they reach a snowflake's surface, the molecules can move about until they find the right slot in the crystal structure. But in DLA the constituent particles of a cluster have no opportunity for such adjustment—they become immobile on contact, and so the way they pack together in the clump is irregular. As a result, the surface of the growing cluster soon becomes very jagged and disorderly. You could say that, because the process is occurring far from equilibrium, the cluster grows too fast for the particles to find the most compact way to pack together, and there are lots of 'packing errors' that get frozen in.

Witten and Sander simulated the DLA process on a computer by placing a particle at the centre of a box and then introducing particles one by one into the box from random points around its edges, allowing them to diffuse until they encounter and stick to any preceding particle. This generates a cluster that grows in tenuous branches (Fig. 2.4). It looks very similar to the structures created in electrodeposition, something that was first recognized by Mitsugu Matsushita of Chuo University in Japan and co-workers in 1984. Brady and Ball proposed that this is because the mechanism of non-equilibrium electrochemical growth shares the same broad features as the DLA model: random diffusion of ions and their instant attachment to the electrode deposit. And that's essentially true, although the details are more complex.

It is clear to see why the random impacts in the DLA model result in very rough-edged clusters. But why are they branched? We could perhaps

FIG. 2.4   A cluster of aggregated particles grown by the diffusion-limited aggregation model. (Image: Thomas Rage and Paul Meakin, University of Oslo.)

imagine instead the formation of a dense mass with a highly irregular border, like a spreading ink blot. Why doesn't this happen?

The answer is that in DLA, as in snowflake growth, small bumps or irregularities are amplified by a growth instability that draws them out into long, narrow fingers. When a bump appears by chance on the cluster surface, it pokes out beyond the rest of the front, so there is a better chance that a randomly diffusing particle will hit it (Fig. 2.5). As a result, the bump grows faster than the rest of the surface. And, crucially, this growth advantage is self-enhancing—the more the bump develops, the greater the chance of new particles striking and sticking to it; in other words, there is positive feedback on the growth of a 'finger'. The probability of acquiring a new particle is highest at the most exposed part—the very tip of the bump—so that growth is accompanied by a sharpening and narrowing of the tip, generating a tendril that itself sprouts random appendages through the same growth instability. Notice that the random, diffusive motion of the particles is essential to all of this. If they were instead all propelled towards the cluster along straight trajectories, like rain falling on a road, the edge of the cluster would just advance uniformly.

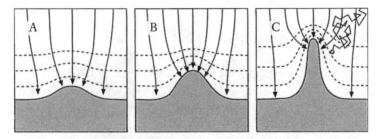

FIG. 2.5 How the DLA model acquires a growth instability that promotes branching. Small protrustions on the aggregate surface accumulate new particles faster than the surrounding flat surface, and so they become increasingly accentuated (a–c). These protrusions will themselves have random irregularities that blossom into new fingers, so the cluster becomes highly branched. Here the dashed lines show the contours of average density of incoming particles, and the solid lines show the average flow of particles, which becomes focused towards branch tips. Each individual particle takes a highly tortuous path—a random walk. One such is depicted in c.

Not only does the 'surface' of the cluster become ramified and diffuse, but the interior never gets 'filled in'. That is because the sprawling branches make it very difficult for a particle to reach very far inside. Given its meandering path, a particle is unlikely to get far down one of the fjords between branches before hitting something—and once it hits, it sticks. This means that the clusters attain open, fluffy shapes. Within the roughly circular envelope of the outermost tips (or spherical in three dimensions), much of the space is empty. Soot particles, which grow by the aggregation of tiny flecks of carbon released in combustion, typically have the same kind of shape, making them as light and airy as soufflé.

WHAT'S IN A BRANCH?

A cluster grown by the DLA mechanism really does resemble a branched electrodeposit, and this might persuade you that the two processes are related. But mere visual similarity cannot constitute scientific proof that there is truly any common mechanism behind their growth. I cannot stress this point too strongly—it is a fundamental issue in the study of pattern formation. When we see two things that look alike, our instinct is to attribute to them the same basic cause—to infer that at root they represent the same phenomenon. I have shown in Books I and II that sometimes this intuition is sound, but sometimes it isn't. There is reason to think that the stripes on a zebra's pelt and the ripples of wind-blown sand stem from the same interplay of activation and inhibition in the creation of the pattern elements. However, stripes in an early-stage fruit fly embryo have a rather different origin. Our facility at recognizing and comparing patterns may help us to identify connections between systems that seem at face value to have no prospect of being related. But this mental faculty is apt to tempt us into making false correlations and untenable analogies.

How do we know when to trust these comparisons and when not? There is no simple answer. I shall show in this book that there *are* remarkable links between many of the different types of branching structure seen in the natural world. But we cannot place much confidence in such claims unless we have, at the very least, some objective, quantitative way of characterizing the patterns we see. How do you measure the shape of a tree or a DLA cluster?

It turns out that even forms as apparently irregular as these have at least one measurable property that is virtually as precise, reproducible, and characteristic as the number of legs on an insect. It is called the *fractal dimension*, and is a measure of how densely packed the branches are. It is not a foolproof means of telling us whether two such shapes are 'the same' in the sense of being formed according to the same principles; but it is pretty accurate at letting us know when they are not. If we measure the fractal dimension of a branched electrodeposit and a DLA cluster and find that these have the same value, we cannot be sure that this is because electrodeposition is basically a kind of DLA process. But if the fractal dimensions are different, we may need to go back to the drawing board to explain how the electrodeposit arises.

There is not really any way of understanding properly what a fractal dimension is without using some maths. But I promised in Book I that any mathematical demands I would make will be minimal, and I do not need to renege on this promise now. Suppose that, instead of forming a branching pattern, a DLA cluster grew so densely that there were no gaps at all between the particles. Then it would be a solid mass whose outer surface simply expands as the cluster gets larger. For clusters growing on a flat surface (that is, in two dimensions), this mass would be roughly circular. In that case, the number of particles in the cluster (N) would increase in proportion to the cluster's area. For a perfect circle, the area is proportional to the square of the radius: to r^2. If, on the other hand, the branches were straight-line chains of particles, so that they formed a many-pointed star like an asterisk, then the number of particles would increase in proportion to the length of each of the arms: in other words, proportional to the radius r of the circular envelope defined by the arm tips. This corresponds to an extremely open, 'empty' cluster. Now, a real DLA cluster like that shown above lies somewhere between these two extremes: N increases as the cluster 'size' r increases, but neither in proportion to r nor to r^2 (that is, to r raised to the power 1 or 2). Rather, N grows in proportion to r raised to a power (exponent) between 1 and 2. This means that the DLA cluster fills up the two-dimensional plane rather more densely than the asterisk-like cluster but less densely than a compact, circular cluster. The exponent, which is the fractal dimension, can be calculated by analysing the density of particles in the cluster, and for two-dimensional DLA clusters it has the value 1.71.

Put this way, the fractal dimension may sound unremarkable—merely a number that falls out of the model. But it in fact implies that there is something odd about a DLA cluster: in a sense, it lies 'between dimensions'. We are used to living in a three-dimensional world, in which objects have 'bulk'—they have a volume, enclosed by surfaces. There are objects in the world that are to all intents one- and two-dimensional too. Laid out straight, a piece of string is one-dimensional: you could say that it has 'length' but no 'width' or 'height'. Of course, it *does* have width and height, but these are negligible in comparison to the length. The piece of string is strictly speaking a three-dimensional object in which two of the dimensions are reduced to almost nothing; if the string were infinitely thin, it would be truly one-dimensional. Likewise, a sheet of paper extends in two dimensions but has negligible extent in the third (the thickness)—it is more or less a two-dimensional object. But a DLA cluster is neither like a piece of string nor like a sheet of paper: it is neither one-dimensional nor two-dimensional, but 1.71-dimensional.

What does that mean? We will look into this question later, but for now it will suffice to say that it challenges the notion that the object has a boundary in any meaningful sense. A piece of paper has edges, and a piece of string has ends: these are the places where the objects stop. But for fractal objects, there is no well-defined place where they 'stop'. If you're right at the 'edge' of such an object, you can never be sure whether you are actually standing inside it or outside, because there is no edge as such. Yes, this is very odd. Just stick with it for a while.

The fractal dimension d_f is a meaningful and useful property of the DLA growth process because it is robust and dependable. It stays the same while the cluster grows bigger and changes shape; and two different DLA clusters, while differing in the precise positions and convolutions of their branches, will have exactly the same value of d_f. In this sense, while we can talk about the 'shape' of a DLA cluster in vague, metaphorical terms, perhaps the only way we can be more precise about this shape is to characterize it by the fractal dimension.

This quantity d_f reflects the rules according to which the cluster was grown. If we change these rules, for example by allowing new particles to make a few short hops around the surface before finally sticking irrevocably, we will very probably obtain a branched cluster with a different value of d_f. Sometimes changes like this will produce very marked changes in the appearance of the clusters—they might develop very stout or very wispy branches, for instance. But the effect of other changes to the growth rules might be rather subtle, so that by visual inspection we will be unable to say whether the clusters are 'the same' or not. The fractal dimension provides a well defined measure by which we can distinguish such differences.

Here's an example. In Fig. 2.6 I show another mineral dendrite, formed from manganese oxide in a fracture plane of a quartz crystal. Is this the same kind of cluster as that in Fig. 2.1? By eye, you probably wouldn't want to place bets. But by calculating its fractal dimension, we can pronounce confidently that the two are different—the earlier dendrite has a fractal dimension of 1.78, whereas for the one shown here it has a value of about 1.51. You can perhaps see that the smaller the fractal dimension, the wispier the cluster.

Branched electrodeposits like that in Fig. 2.3a commonly have a fractal dimension of about 1.7,* and this can give us confidence that their

---

*In three-dimensional growth a DLA cluster has a fractal dimension of about 2.5, while Brady and Ball showed that electrodeposits grown in three dimensions have a fractal dimension of around 2.43.

FIG. 2.6   A mineral dendrite inside a quartz crystal. (Photo: Tamás Vicsek, Eötvös Loránd University, Budapest.)

mechanism of formation does share something in common with the DLA process. But what about mineral dendrites? You might have guessed from their shapes alone that the DLA model offers a good description of their formation too, but we now discover that two mineral dendrites can have fractal dimensions that differ not only from that of a DLA cluster but from one another.

The French physicist Bastien Chopard and his colleagues have shown that the formation of mineral dendrites can in fact be explained by a more sophisticated version of DLA. In their model, the solution of ions that form the mineral dendrite diffuses through cracks in the surrounding rock, and the ions undergo a chemical reaction when they encounter each other. The dendrite is formed of metal and oxide ions: crudely speaking, we can imagine these two dissolved ions diffusing until they meet and form an insoluble black deposit. This is emulated in the model by positing two soluble chemical species A and B that move through the surrounding medium at random and may react to form a dissolved compound C. If enough C accumulates in a particular region, the solution becomes supersaturated and C precipitates in the form of a dark material D, which then stays put. If, meanwhile, a single C particle encounters a cluster of D, it

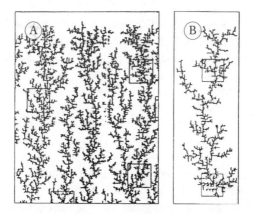

FIG. 2.7 Mineral dendrite patterns generated by a computer model in which particles diffuse by random walks. When particles encounter one another, they react to form a dark deposit. The model can generate forms with a range of fractal dimensions: shown here are examples with a fractal dimension of 1.78 (a), similar to the mineral dendrite in Fig. 2.1, and 1.58 (b), close to that in Fig. 2.6. (The square boxes show the regions selected for calculating the fractal dimension.) (From Chopard et al., 1991.)

too will precipitate, becoming stuck to the cluster. Although couched in different terms, this is actually a *reaction–diffusion* model akin to those I described in Book I, where we saw that they are fertile sources of pattern.

Chopard and colleagues found that their model generates fractal clusters much like real mineral dendrites (Fig. 2.7). The fractal dimension of the model clusters varies according to the concentration of species B: the researchers were able to generate simulated mineral dendrites with fractal dimensions of 1.75 and 1.58 (close to the values for the natural samples I have shown) by changing this concentration.

FRACTALS EVERYWHERE?

I can think of no better illustration than fractals of the fact that science, like any other human activity, is prey to fashion. In the 1980s fractals were the thing: they were celebrated in popular books, in posters and postcards and on T-shirts. The most famous fractal object was the abstract structure discovered in the 1970s by mathematician Benoit Mandelbrot, working at IBM's research centre in Yorktown Heights, New York: a bulbous black kidney shape now called Mandelbrot set (Fig. 2.8). This beast, decorated with wispy filaments and often sporting furiously (and spuriously) coloured spirals, erupts across the mathematical plane in response to a deceptively simple algebraic equation. Mandelbrot pioneered the recognition of fractals as a new type of geometry that is abundantly expressed in nature (there were antecedents which had hinted at this idea). He coined the word 'fractal' in 1975, and in his seminal book *The Fractal Geometry of Nature* two years later he showed how fractal forms may be found throughout the natural world, from coastlines to the shapes of plants and clouds (Plate 2). By the 1980s computer-generated fractal landscapes and imitations of organic complexity had found their way into films and commercials. Today these elaborate, convoluted

FIG. 2.8 The Mandelbrot set, a mathematical fractal defined by the boundary between the 'basins of attraction' for solutions to an equation. The solutions lie on a flat plane, with real numbers running vertically and 'imaginary' numbers horizontally.

forms no longer possess the allure of novelty; in scientific research, where once it was a matter of interest merely to identify a new fractal structure, this now elicits a weary shrug of the shoulders, for scientists have grown accustomed to the notion that they are ubiquitous.

What are fractals? Mandelbrot called them 'monsters', for they have properties that are strange, puzzling and, to a mathematician, even a little frightening. Mandelbrot asserted that fractals lie in the middle ground between forms that display the familiar, regular geometric order that we associate with Euclidian mathematics—simple shapes such as triangles, squares, and circles—and those that seem merely random and chaotic. They are, he said, a kind of orderly, geometric chaos: an apt collision of contradictory terms.

If you look closely at the edges of the Mandelbrot set, you will see that it is bordered with branching tendrils that may make you think of mineral dendrites and DLA clusters. The link between these structures can be made more explicit and formal. What both the Mandelbrot set and DLA clusters, along with all other breeds of fractal structure, have in common is that they look more or less unchanged as you peer at them more and more closely, zooming in as though ramping up the magnifying power of a microscope. Take a patch of a DLA cluster and magnify it, and you will see a ramified object that looks much the same as the whole. Do the same again to a patch of the magnified region, and you find the same thing (Fig. 2.9). What this means is that you cannot tell the degree of

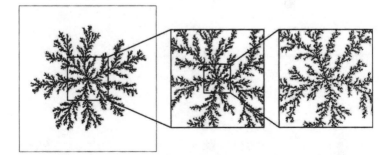

FIG. 2.9   When a DLA cluster is magnified, it looks more or less the same at each scale. This property is called self-similarity, or scale invariance. (Image: Thomas Rage and Paul Meakin, University of Oslo.)

magnification purely from appearance, because there is nothing to guide you. In contrast, you can quite easily assess the scale of an aerial photo of a town because there will be features, such as cars, houses, roads, that provide a known measure of length. (Because, as we shall see, many geological structures are fractal, geologists often leave a hammer in photographs of rock faces to provide a reference scale.) This property of keeping the same form under different levels of magnification, which is to say, under changes of scale, is called *scale invariance*, or more loosely, *self-similarity*. One aspect of the self-similarity of the Mandelbrot set is that, as you zoom in on any region of the perimeter, the ominous black bulb keeps reappearing like a malformed Russian doll.

Because of scale invariance, fractal forms have no boundary. There are points in the plane of the Mandelbrot set that are unambiguously inside or outside the black region, but if you are right on the 'edge' then you cannot be sure which side you're on: each time you zoom in further, you see more of the convolutions. This can be illustrated more clearly with reference to a well-known fractal form called the Koch snowflake, made by repeatedly kinking a boundary line at ever smaller scales. Take a straight line and introduce an equilateral kink in the middle third (Fig. 2.10). Now repeat this process for each of the straight-line segments that result, and go on doing so again and again, each time at a smaller scale. You end up with a line that zigs and zags somewhat like the repeatedly and symmetrically branching arm of a snowflake. At each step, you can say precisely where the boundary line is. But on the next

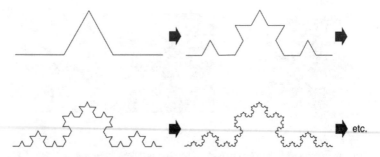

FIG. 2.10 A fractal object called the Koch snowflake is produced by introducing identical kinks into a line at successively smaller scales.

step, that point in the plane may have been engulfed by a new kink. If we continue this process an infinite number of times, the line is infinitely convoluted and it is impossible to say exactly where it passes. The 'Koch snowflake' remains a finite shape, but its edge has become fuzzy.

The Mandelbrot set delights visual artists because it is a literally endless source of baroque patterns that stimulate the eye. It is a kind of map drawn on a mathematical plane in which each point represents a number. All the numbers within the black boundary of the set are associated with one solution of the equation that defines the Mandelbrot set, while all the numbers outside are associated with another solution.* The technicoloured swirls commonly depicted in these maps encode how quickly the numbers they encompass get converted into one of these solutions. The details don't matter; what is remarkable is the way the Mandelbrot set generates patterns and forms that have a hundred echoes: ferns, vortices, lightning. Fractal geometry is inherently visual. Mandelbrot says that the nineteenth-century French mathematicians Lagrange and Laplace:

> once boasted of the absence of any pictures in their works, and their lead was followed almost universally. Fractal geometry is a reaction against the tide, and a first reason to appreciate fractal geometry, because of the 'characters' it adds to the 'alphabet' Galileo had inherited from Euclid, [is that they] often happen to be intrinsically attractive. Many have promptly been accepted as works of a new form of art. Some are 'representational', in fact are surprisingly realistic 'forgeries' of mountains, clouds or trees, while others are totally unreal and abstract. Yet all strike almost everyone in forceful, almost sensual, fashion.

---

*'Associated with' is a little vague. What I mean by it is the following. The 'Mandelbrot equation' is essentially a prescription for generating one number (call it z_2) from another (z_1): the prescription is $z_2 = z_1{}^2 + c$, where c is a constant. You begin with $z_1 = 0$ and calculate z_2, and then use *that* number as z_1 and calculate a new z_2, and so on. Eventually you find that your z_2 either heads towards infinity or it doesn't, depending on the value you choose for c. The Mandelbrot set is the set of all numbers c that don't give you an infinite solution as you keep iterating the equation. The reason the Mandelbrot set is two-dimensional is that c has both a real and an imaginary part—real numbers lie along the horizontal axis, imaginary ones along the vertical axis. Imaginary numbers are multiples of the square root of minus 1, but if this is unfamiliar, it needn't detain us a moment longer.

It must be said that this is a double-edged virtue. Some mathematicians sniffily dismissed the fad for fractals as a kind of fancy computer-graphic doodling, nothing to do with their precise and intricate art of numbers. And I think it must be said that the 'art' that emerges from fractals generally does not seem very far away from the tiresome kitsch of rock album covers and psychedelia.*

But there are other reasons to be wary of making fractal geometry the key to natural form. Once you get the hang of what a fractal structure looks like, you imagine you see them everywhere. Without doubt the branched fractals typified by DLA clusters do represent one of nature's universal forms, and are splendid examples of how complex, 'life-like' shapes can be the product of relatively simple and entirely unbiogenic processes. But we must remember that not all branched patterns are fractal; and perhaps more importantly (since researchers had, in the first flush of fractal frenzy, a tendency to forget it), just saying that a structure is fractal doesn't bring you any closer to understanding how it forms. There is not a unique fractal-forming process, nor a uniquely fractal kind of pattern. The fractal dimension can be a useful measure for classifying self-similar structures, but does not necessarily represent a magic key to deeper understanding.

Furthermore, what we mean by 'fractal' in the natural world is by no means clear. We saw above that the defining property of fractals—the feature that gives them a fractal dimension—is self-similarity, or invariance under different levels of magnification. For the Mandelbrot set, this invariance continues however closely we look. The same applies to the theoretical Koch snowflake. But, of course, for real objects the fine details have to stop somewhere, if only because eventually we reach the scale of atoms. In fact, for most 'real' fractals the self-similarity stops well before that. For a branched metal electrodeposit we have seen that the building blocks are tiny crystals made up of millions of atoms: at the scale of these crystallites the surface of the deposit becomes flat, corresponding to a crystal facet. And for many of the popular examples of 'natural fractals', such as ferns, there are only a few self-similar repetitions of the branching structure. Below the scale of the smallest leaflets, the leaf structure is no longer fractal, no longer

*I have discussed the aesthetics of 'fractal landscapes' in *Nature* 438 (2005): 915.

between one- and two-dimensional, but fills up space entirely: the fern becomes a fully two-dimensional object. Snowflakes are similar: typically their six arms have side-branches, and there the branching stops.

Many researchers suggest that, while natural fractals clearly cannot be truly fractal at all size scales, to qualify for the designation they need at least to be self-similar over size scales of several factors of ten—say, for magnifications of up to a thousandfold. But David Avnir of the Hebrew University of Jerusalem in Israel and his co-workers have shown that most real objects declared 'fractal' tend to lose their self-similarity at little more than ten times magnification, or a hundred times at best. This often means that the object in question simply has a rough surface. Nature's geometry is fractal, the researchers say, only because people have become content to call such irregularity 'fractal' and not because it corresponds in any meaningful way to Mandelbrot's original definition of mathematical fractals in all its recursive infinities.

That's as may be. But for highly branched objects like mineral dendrites and DLA clusters, it is quite possible to ascribe a meaningful fractal dimension so long as we view the objects at scales where the high degree of branching is evident. In this case, the fractal dimension provides a good way of making comparisons between patterns that 'look' alike. And I shall continue to use the word 'fractal' to describe them.

SQUEEZE PATTERNS

The Spanish physicist Juan Manuel Garcia-Ruiz was assailed by fractal branches even as he sat down for a quiet cup of coffee in the Hotel Los Lebreros in Seville. Across the broad plate-glass windows of the coffee shop crept three of Mandelbrot's monsters, like ghostly plants growing within the glass (Fig. 2.11). Each window pane was made from three laminated glass sheets, separated from one another by thin plastic films. The laminates were imperfectly sealed at their edges, so air could find its way between the sheets. In the Spanish heat, the plastic had become soft and viscous, and the air had pushed its way through the film in a bubble whose advancing front broke up into a fine tracery of fingered branches, just like the tendrils of DLA clusters. Recognizing this characteristic

FIG. 2.11 Viscous fingering patterns in the layered window pane of the Hotel Los Lobreros in Seville. These patterns are formed as air penetrates into the plastic film between the glass planes. (Photo: Juan Manuel Garcia-Ruiz, University of Granada.)

pattern, Garcia-Ruiz took photos and analysed them. He found that the fractal dimension of the bubble was about 1.7.

The process in which air forces its way under pressure into a viscous medium as a branching bubble is called viscous fingering. It has been studied a great deal, because it is relevant to some very practical problems in engineering. For instance, oil is often extracted from oil fields by injecting water through a borehole into the oil-saturated porous rock. The idea is that the water, which does not mix with oil, should advance in a front that pushes the oil to the wells at the edge of the field. But if viscous fingering occurs, the water front breaks up into narrow fingers and the efficiency with which oil is displaced and recovered is very low.

Does this sound familiar? Viscous fingering is essentially the same as the process described by Jean-Jacques Scheuchzer in the early eighteenth century as an analogue of how mineral dendrites form (page 30). In Scheuchzer's process the air was not forced into the liquid under pressure but was pulled in by the vacuum created when two plates separated by a liquid film are pulled apart. That, however, is basically the same thing.

It is no coincidence that viscous fingering produces forms similar to those made in DLA, even though at face value the phenomena seem quite different—for, like DLA, viscous fingering involves an instability that makes bulges at the interface prone to elongation. We have seen that snowflakes and other dendrites are also governed by this kind of effect, in the form of the Mullins–Sekerka instability. For viscous fingering, the origin of the branching instability was identified in 1958 by P. G. Saffman and Geoffrey Taylor, who studied viscous fingering using an apparatus devised by a nineteenth-century British naval engineer named Henry Hele-Shaw. He was aiming to study how water flows around a ship's hull, but his equipment, now called a Hele-Shaw cell, has now become a standard tool for research into branching patterns. It consists of two horizontal, parallel plates with a narrow gap between them. The top plate is made of some transparent material and has a hole in the centre, through which a relatively inviscid fluid (such as air or water) can be injected into a more viscous one (such as glycerine or an oil) held between the plates (see Appendix 1).

The edge of the air bubble (say) moves forward into the oil because the pressure in the air just behind the interface is greater than that in the oil just in front of it. The speed at which the interface advances depends on how steep this pressure gradient is. This is entirely analogous to the role of the temperature gradient in the Mullins–Sekerka instability; and by the same token, the gradient is steeper at a bulge, rising to its highest value at the tip. So again there is a self-amplifying process in which a small bump formed at random at the interface advances faster and becomes more drawn-out. This is the so-called Saffman–Taylor instability.*

The analogy between viscous fingering and DLA carries through to the mathematical level: the equations describing them are equivalent. Both these and the Mullins–Sekerka instability correspond to a set of equations derived from the work of the eighteenth-century French mathematician and scientist Pierre-Simon Laplace, which earns them all the generic name of Laplacian instabilities. Perhaps if Laplace had been able to think about maths more pictorially, he would have discovered the origin of these fractal branching patterns nearly two centuries earlier.

However, tenuous fractal patterns resembling those of DLA occur in viscous fingering only under rather unusual conditions. More commonly one finds a subtly altered kind of branching structure: the basic pattern or

*In view of Scheuchzer's early work, and its elaboration by the Abbé de Sauvages in 1745 (who proposed Hele-Shaw's arrangement of a liquid injected into a second, more viscous one), the French physicist Vincent Fleury has proposed that this be renamed the Scheuchzer–Sauvages instability.

'backbone' of the network has a comparable, disorderly form, but the branches themselves are fat fingers, not wispy tendrils (you can see this in Fig. 2.11). And under some conditions the bubbles cease to have the ragged DLA-like form at all, instead advancing in broad fingers that split at their tips (Fig. 2.12a). This sort of branching pattern is called the *dense-branching morphology*, and it is not really fractal at all: it fills up the available space almost entirely, and so has a fractal dimension close to 2. Why, if the same basic tip-growth instability operates in both viscous fingering and DLA, do these different patterns result?

All viscous-fingering patterns differ from DLA in at least one important respect: they have a characteristic *size scale*, defined by the average width of the fingers. This can be seen most clearly for low injection pressures, when the bubble front advances slowly with an almost regular undulation around the perimeter (Fig. 2.12b). In other words, the pattern now has a particular wavelength or scale, and so it is no longer a scale-invariant fractal.

This scale arises for the same reason that the branches in snowflake-like dendritic growth have a particular width: surface tension makes it costly to form an interface, limiting the minimum size of the crenellations. The fat fingers represent a compromise between the Saffman–Taylor instability, which favours the growth of branches on all length scales, and the

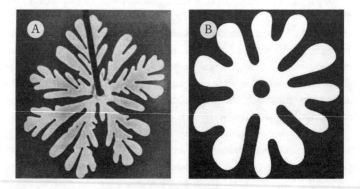

FIG. 2.12   The dense-branching morphology in the Hele-Shaw cell (*a*). This is only marginally fractal, with all fractal dimension close to 2. At low injection pressures, the advancing bubble has a rather smoothly undulating edge with a fairly well-defined wavelength (*b*). This is no longer a fractal shape at all (Photo: *a*, Eshel Ben-Jacob, Tel Aviv University.)

smoothing effect of surface tension, which washes out bulges smaller than a certain limit. In the DLA model, on the other hand, the cluster has essentially no surface tension, and so the branches may become as thin as they can be, equal to the width of the aggregating particles themselves. A wispy DLA-like 'bubble' can be produced experimentally in the Hele-Shaw cell by using fluids whose interface has a very low surface tension. Alternatively, this can be achieved by making the growth of the bubble more 'noisy'—more prone to random fluctuations that encourage the appearance of new branches at the bubble front. A simple way of doing this is to score grooves at random into one of the cell plates until it is criss-crossed by a dense network of disorderly lines (Fig. 2.13a). If, on the other hand, these grooves are arranged in an orderly grid, that imposes an underlying symmetry on the bubble. For a hexagonal, honeycomb grid, the resulting bubble has a snowflake-like shape (Fig. 2.13b).

This is the attraction of the Hele-Shaw cell: it provides a tool for investigating how a diverse family of branching patterns—DLA-like, viscous fingering, the dense-branching morphology and dendritic snowflakes—are related. A bit of noise tips the pattern one way; a dose of symmetry does something else. The branches are the result of Laplacian growth instabilities that amplify small wobbles of the surface;

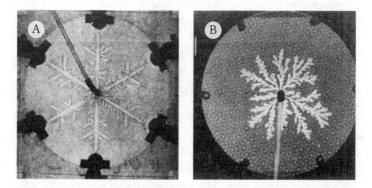

FIG. 2.13   In a Hele-Shaw cell in which the bottom plate is 'randomized' with a criss-crossing web of grooves, the bubble has a shape akin to a cluster grown by diffusion-limited aggregation (a). If the grooves form a regular honeycomb lattice, on the other hand, the bubble resembles a snowflake (b). (Photos: Eshel Ben-Jacob, Tel Aviv University.)

surface tension can moderate that tendency. The results may be fractal, but are not necessarily so. These 'trees' are a subtle blend of enthusiasm and restraint operating at the border of order and chaos.

LIFE IN THE COLONIES

In defiance of our long-standing intuition, then, there need be nothing 'organic' about branching structures. Yet, in a curious inversion, the ideas that have been developed to explain branches in minerals and bubbles are now being exported back into biology, to account for the complex shapes of communities of simple organisms such as bacteria.

Forms with ruffled, frond-like boundaries that radiate from a central focus are familiar in the microbiological world: we see them, for instance, in the way lichen spreads across rock (Fig. 2.14a). These fringed blobs are also found in colonies of malignant cells that grow into tumours (Figure 2.14b). The shape as a whole is not a fractal, but the boundary is: whereas the smooth perimeter of a circle is one-dimensional, the tumour growths have roughly 1.5-dimensional edges. This kind of shape was first described mathematically in 1961 by the mathematician M. Eden, in one of the first examples of a computer model of biological growth. Eden was seeking to model how tumours develop, but any growth process that leads to such wavy-edged, dense shapes is now said to be Eden-like. In Eden's model, particles (cells, say) multiplied on a regular grid, with one cell per grid site. New particles were added at random to sites adjacent to ones already occupied, so that the cluster was constantly accumulating new particles around its perimeter. This is rather similar to Witten and Sander's DLA model, and in fact they acknowledged this debt to Eden. But in DLA the particles drift onto the cluster from the outside, rather than being effectively 'pushed out' from the surface, which makes the clusters more wispy and less compact.

In the late 1980s, Mitsugu Matsushita and H. Fujikawa at Chuo University in Japan saw Eden-like shapes in colonies of *Bacillus subtilis* cultured in flat, circular Petri dishes containing a water-saturated gel made of a substance called agar (Fig. 2.15a). They injected a few bacteria into the centre of the dish, added some of the nutrients needed for growth, and let nature take its course.

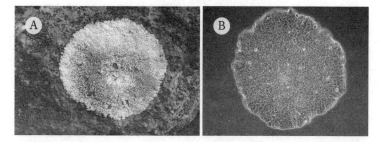

FIG. 2.14   The growth of lichen colonies (a) and of some colonies of tumour cells (b) have a compact, roughly circular shape with a ragged, fractal fringe. This is called Eden growth. (Photos: a, Ottmar Liebert; b, Antonio Bru, Universidad Complutense, Madrid.)

By varying the conditions of growth, they found that they could obtain colonies with very different shapes. The bacteria cannot penetrate into the gel, and so the colony has to push back the gel boundary as it grows. This is comparable to the injection of a fluid into a Hele-Shaw cell, and the 'injection pressure' can be varied by changing the concentration of agar in the growth medium: the more there is, the harder the gel, which can vary in consistency from jelly-like to rubbery. And the harder it is, the greater the resistance to growth of the colony. The shape of the colony also depends on the amount of nutrient available to the bacteria.

Below a certain nutrient concentration, the pattern switches from Eden-like to a now-familiar fractal form (Fig. 2.15b), reminiscent of DLA clusters and sharing with them the same fractal dimension of about 1.7. If, meanwhile, the gel is made softer at these low nutrient levels, the pattern changes from DLA-like to one that more closely resembles the dense-branching morphology (Fig. 2.15c). If there is plenty of nutrient and the gel is soft, the colony expands in a dense mass with a roughly circular perimeter.

The growth of a bacterial colony is clearly much more complex than the aggregation of inanimate particles or the expansion of a bubble. The cells eat, they replicate, and they may move around. Yet the growth patterns that result are the same. It seems that this process, too, is governed by branching instabilities.

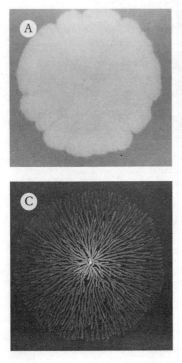

FIG. 2.15 Bacterial colonies in a plate of agar gel can grow into several different shapes depending on the growth conditions, such as the availability of nutrients and the hardness of the gel. Here are several examples in *Bacillus subtilis*: Eden growth (*a*), DLA-like growth (*b*) and the dense-branching morphology (*c*). (Photos: Mitsugu Matsushita, Chuo University.)

But there are important differences. The Japanese researchers found, for example, that if the gel gets too hard then the bacteria simply cannot move. Under a microscope, they could see that bacteria in the dense-branching colonies were swarming about, while those in the DLA-like and Eden-like colonies just sat there. If they grew colonies of *Bacillus* mutants that lacked the whip-like appendages which enable them to swim around, only the DLA and Eden patterns were formed no matter how soft the gel was.

In Book II we saw how models of collective motion can explain the coherent swarming behaviour of organisms ranging from single cells to fish and birds. One of the first of these models was devised by the Hungarian physicists Tamas Vicsek and András Czirók, who showed that complex behaviour can result when the individual organisms follow a set of rather simple rules dictating how their actions are influenced by those of their neighbours. In the mid-1990s they teamed up with Eshel Ben-Jacob and his co-workers at Tel Aviv University to apply similar ideas to the growth of bacterial colonies.

They assumed that the most significant facts to include in a model like this are that the bacteria move, feed and reproduce. So they adopted the following rules:

1. The bacteria move at random.

2. While food is available, the bacteria feed at a steady rate.

3. If they eat enough, they reproduce (one cell splits into two); if the food runs out, they stop moving.

4. The dispersal of food (nutrient) throughout the system takes place by diffusion.

A single colony might contain as many as ten billion individual 'particles' (cells), which is far too many for a computer to cope with. The researchers therefore grouped cells together into 'walkers', each containing many thousands of cells, and assumed that these walkers moved around according to the same rules. Thus the walkers, not individual cells, were the fundamental particles of the model. They moved about on a regular underlying grid, and the boundary of the colony could advance onto a new grid point only when that point had been struck a certain number of times by the moving walkers. By varying this number, the researchers could simulate the effect of making the gel harder.

With nothing more than these elements, they found that their computer model of the growing colony produced the DLA-like and dense-branching patterns seen in the experiments. The lower the concentration of food, the more tenuous the branches become (Fig. 2.16).

But in the experiments a curious thing happens when the amount of nutrient is very low: the colony suddenly becomes denser again. This does not happen in the model—the branches just go on getting thinner as food becomes scarcer. The researchers figured that it is at this point, when things look really desperate, that the starved bacteria start to do something only living 'particles' can do: talk to each other. As we saw in Book I, the language of bacteria is a chemical one: they communicate by emitting chemicals which then guide the cells' direction of motion in a process known as chemotaxis. This enables *B. subtilis* to aggregate into clumps, where the hitherto identical cells take on distinct roles, rather as though they have become a multicellular organism. Some become spores, which remain in suspended animation until conditions are more favourable.

So the researchers added to their model a simple description of chemotaxis. But they made the chemical signal a *repellant* rather than an attractant: any walkers that encounter it have a tendency to wander *away* from the source. Cells were assumed to emit the chemo-repellant if they became immobile due to lack of nutrients. This addition to the model produced the denser branching patterns at very low nutrient levels (Fig. 2.17), as seen experimentally. There is, however, no evidence that *Bacillus subtilis* really employs repulsive chemotaxis, so one cannot yet be sure that this apparent success of the model is not just a happy coincidence.

INVASION OF THE MUTANTS

Ben-Jacob and his co-workers not only studied bacterial growth theoretically; they also learnt the microbiological techniques needed to grow

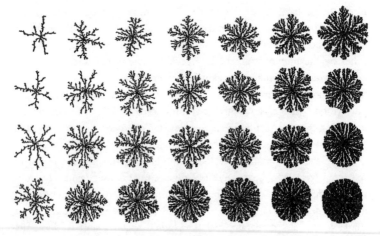

FIG. 2.16 A computer model of bacterial growth that applies only a few simple rules to the movement and multiplication of cells generates DLA-like branching patterns that become increasingly sparse as the gel medium becomes harder (bottom to top) or as the nutrient concentration decreases (right to left). (Image: Eshel Ben-Jacob.)

their own colonies of *Bacillus subtilis*. While some of their experiments yielded the same kind of growth patterns as Matsushita and Fujikawa had seen, they also found some new ones. Occasionally, a colony that was steadily advancing in one pattern would suddenly sprout an entirely different kind of sub-colony (Fig. 2.18). If cells from such an outgrowth were extracted and used as the seeds of a new colony, this would exhibit the new pattern, too. It was a mutant colony.

For unlike aggregating metal atoms or smoke particles, bacteria can mutate: random errors in duplicating a cell's DNA when it divides spawn genetic variations in the progeny. These mutations happen all the time; some are fatal, others have no observable effect. But just occasionally a

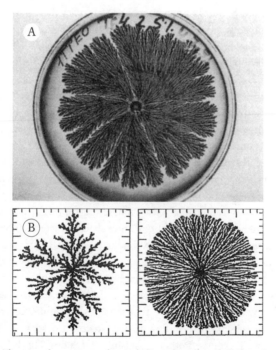

FIG. 2.17 There is a change in growth morphology of *Bacillus subtilis* from the DLA-type pattern (Fig. 2.15*b*) to a denser branching pattern (*a*) at low nutrient levels. When chemotaxis is included in the computer model, it reproduces this switch (*b*: the image on the right is obtained for a lower nutrient concentration than that on the left). (Photo and images: Eshel Ben-Jacob.)

FIG. 2.18 A mutant colony with a new growth pattern sprouting from a dense cluster of *Bacillus subtilis*. (Photo: Eshel Ben-Jacob.)

mutation will make the new cell better adapted, more able to replicate efficiently, than the parent cell. This is exactly how Darwinian natural selection works.

Ben-Jacob and his colleagues suggested that what they were seeing in these spontaneous changes of growth pattern was natural selection in a Petri dish. The explosive growth and dominance of the new pattern signalled the superior fitness of the mutants—although whether the new pattern was a cause of this or an incidental side-effect was not clear.

This process supplied a variety of new forms. When a mutant colony emerged, the researchers would breed its cells to obtain a new strain of *Bacillus* with new pattern-forming behaviour. Some of the mutant patterns were familiar: dense-branching colonies, for example, which Ben-Jacob and colleagues called the tip-splitting or T morphotype. But other mutant patterns were unlike anything seen in non-living systems. One consisted of elegant hook-like twists that all curved in the same direction, creating a colony reminiscent of a Chinese dragon (Fig. 2.19*a* and Plate 3*a*). They dubbed this the chiral or C morphotype ('chiral' derives from the Greek word for hand, as these hooks can twist either in a left- or right-handed direction). Meanwhile, a mutant they called the vortex or V morphotype advanced as mobile, roughly circular droplets of cells that left tendrils in their wake (Figure 2.19*b* and Plate 3*b*). Under the microscope, the researchers could see that the cells in the droplets were all rotating in a spiral vortex—a flow pattern that, as I describe in Book II, has been seen in groups of other organisms such as fish. This behaviour is not unprecedented in bacteria: in 1916 the microbiologist W. W. Ford reported such vortices in *Bacillus* colonies that were consequently given

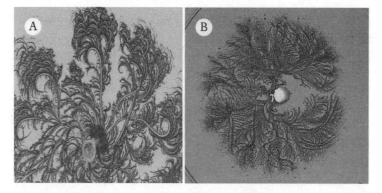

FIG. 2.19 New growth patterns of *Bacillus* bacteria: the chiral (*a*) and vortex (*b*) modes. (Photos: Eshel Ben-Jacob and Kinneret Ben Knaan, Tel Aviv University.)

the species name *circulans*. Apparently the vortex mutants of *B. subtilis* have acquired a similar kind of motion.

The researchers have been able to devise models of cell motion that reproduce these patterns, but it remains very difficult to prove that such models really capture the right biological ingredients, rather than just generating a coincidental similarity of form. The biochemist Jim Cowan has some harsh but reasonable words to say about people who attempt to develop simple models of complex systems like this: 'They say "Look, isn't this reminiscent of a biological or physical phenomenon!" They jump in right away as if it's a decent model for the phenomenon, and usually of course it's just got some accidental features that make it look like something.' It is as well never to lose sight of this brand of scepticism, which insists that just because we can make a pattern on a computer or in a theory, that doesn't mean nature weaves it using the same rules.

URBAN SPRAWL

Bacteria are not, of course, the only organisms that grow in colonies which depend on food and other resources and which suffer if the population becomes too dense. For many of us, this sounds a little bit like home.

Indeed, the notion of city as organism is an old one. To Lewis Mumford, whose 1938 book *The Culture of Cities* was for a long time the bible of progressive urban planners, 'the growth of a great city is amoeboid ... the big city continues to grow by breaking through the edges and accepting its sprawl and shapelessness as an inevitable by-product of its physical immensity'. The American urban theorist Jane Jacobs insisted in her analysis of the decline of US cities in the 1950s that they should be considered as living organisms with their own metabolism and modes of growth, a notion that led her to become one of the first people to recognize how complex systems of many interacting parts can display orderly, self-organized behaviour.

It takes an eye receptive to the character of organic form to notice this, for it is all too easy to regard the city as an amorphous chaos (Fig. 2.20*a*). In this fragmented, irregular cluster of little units, it is hard to discern any sign of the regularity that urban planners might try to impose. Instead, this structure is reminiscent of that one sees when small particles aggregate at random (Fig. 2.20*b*), rather like the clumping of dust and soot or the flocculation of river silt. As we have now seen, however, such processes do generate characteristic forms, with boundaries that are more or less branched and ramified, and which may be analysed mathematically using the techniques of fractal geometry.

With this in mind, the British geographers Michael Batty and Paul Longley have asked whether models based on diffusion-limited aggregation—the source of fractal, branching clusters—can tell us anything about the growth and form of cities. This was a radical departure from tradition. Since the major preoccupation of urban planners is with the design of cities, they have generally attempted to analyse city forms with those efforts in mind. And so their theories have tended to focus on cities in whose outlines the guiding hand of human design is clearly discernible. But hardly any cities are like this. In spite of the efforts of planners to impose a simplistic order, most large cities present an apparently disordered, irregular scatter of developed space, in which residential neighbourhoods, business districts, and green spaces are mixed haphazardly. By focusing on centres where planning has created some regularity (like the US grid-iron street plan), urban theorists have often ignored the fact that cities tend to grow organically, not through the dictates of planners.

The planned, geometrically ordered city has long been seen as the ideal. As geometry became a dominant aspect of ancient Greek thought, its influence extended beyond architecture into the way in which the buildings themselves were arranged in settlements. The grid street plan was evident even in the cities of Babylon and Assyria, but is most apparent in the towns built by imperial Rome: it was a scheme that allowed these settlements, often beginning as military encampments, to be erected quickly.

Another favourite scheme of geometrical planners was the radial or circular city plan, in which main thoroughfares radiate from a central hub like the spokes of a wheel. This became a popular motif during the rationalistic climate of the Renaissance and the Enlightenment. Christopher Wren imagined London rebuilt after the Great Fire of 1666 as a grid connecting radial centres, one of them focused on his new cathedral of St Paul's. But he never saw it realized, because cities that have grown dishevelled will not tolerate an imposed order: the jumble of streets in the old city reasserted itself faster than Wren or anyone else could construct new ones.

The fact is that cities are not static objects but growth forms with a logic that eludes our rectilinear geometric tradition. They are structures that emerge out of equilibrium. Planners and urban theorists are still coming to terms with this fact, and they view it with some ambivalence: is it a

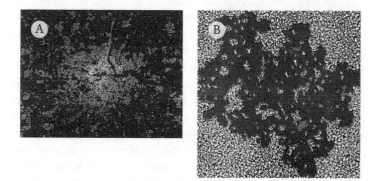

FIG. 2.20   The shape of a city like London, shown here as a map of employment density (a), is highly irregular. Rather similar shapes can be seen in the aggregation of microscopic plastic spheres suspended in a liquid (b). (Images: a, Michael Batty, University College, London; b, Arne Skjeltorp, Institute for Energy Technology, Kjeller.)

good or a bad thing for cities to evolve this way? Should they be allowed to grow 'naturally', or should we try to impose some structure on it all? Does irregular growth mean that cities will descend into chaos, spawn slums, and lose control of public services? Or do they grow as they 'need' to, finding their own optimal paths and solutions, so that attempts to enforce regularity just create inefficiencies and sterile, unneighbourly living spaces?

There is surely no universal answer, not least because it is likely to depend on the social and economic context in which urban growth occurs. But before we can even assess the issues, we need to have a description of *how* cities grow. That is what bothered Batty and Longley—they felt that this description was lacking. As a result, it was impossible to predict what a city might look like five years hence, hampering the ability of planners to anticipate the likely requirements for transportation, water, power supplies, communications networks, and so forth. 'There is a need', Batty and Longley wrote in their 1994 book *Fractal Cities*, 'for a geometry that grapples directly with the notion that most cities display organic or natural growth, that form cannot be properly described, let alone explained, using Euclidean geometry'. The title is a giveaway: Batty believed that the appropriate new geometry was to be found in the concept of fractals developed by Benoit Mandelbrot—the 'geometry of nature'.

For decades, urban theorists have known that the structure of cities can be described by mathematical relationships called power laws (see page 38). For example, the population density often decreases fairly steadily as one moves outwards from the city centre, typically by some relationship in which this density is proportional to the inverse of the distance raised to some power. A similar relationship describes how the number of settlements (cities, towns, villages, hamlets) in an urbanized area depends on their size (in population or area, say): there are many more small villages than there are towns, and still fewer cities, and the power law quantifies that fact. Planners and geographers could measure these relationships, but they could not work out why they arose from the underlying economic and demographic processes that determine the evolution of an urban area. They didn't know the natural rules of urban growth.

In the early 1990s, Batty and others used the methods of fractal analysis to deduce the fractal dimensions of cities. As we can see from Fig. 2.20*a*, the boundaries of cities tend to be irregular and fragmented, and we might imagine that they are indeed fractal objects, extending over a two-dimensional space but not fully filling it. The studies showed that in fact the fractal dimensions of cities span a wide range, typically between about 1.4 and 1.9. London in 1962 has a fractal dimension of 1.77, for Berlin in 1945 it is 1.69 and for Pittsburgh in 1990 it is 1.78. In general a city's fractal dimension increases slowly over time, reflecting the fact that more and more of the 'free' space between centres of development tends to get filled in, making them increasingly two-dimensional. Fractality extends to the characteristic networks of urbanization too, such as transport or power lines. The transport networks of Lyons, Paris and Stuttgart, for example, are branched fractals with dimensions ranging from only just over 1 (a very sparse network) to almost 1.9. The Paris metro and suburban rail network, for instance, has a fractal dimension of 1.47 (Fig. 2.21).

These numbers mean little by themselves. The challenge is to understand how they come about: to look for a model of a growth process that reproduces them. Batty and Longley realized that the mean fractal dimension of the cities that they and others had analysed (about 1.7) is rather close to the fractal dimension of clusters formed by diffusion-limited aggregation (1.71). So, as a start, they decided to mimic urban growth using the DLA model. Recall that in this model particles execute random walks until they strike the perimeter of a growing cluster, whereupon they stick where they hit. Batty and Longley suggested that something similar

FIG. 2.21 The Paris Metro is a branched network with a fractal form. (Image: M. Daoud, CEN Saclay.)

happens as cities grow: new development units, such as business or residential neighbourhoods, are gradually added to the city with a probability that is greater at the city's perimeter, since there is more space there for development. Of course, this highly simplistic model ignores a great deal that is important for urban development, not least all efforts of planners to impose some order on it. Yet the researchers found that some of the growth laws observed in real cities are similar to those that apply to DLA clusters.

Even so, DLA clusters and fractal cities do not look much alike: cities are typically denser, more compact. This shape can be more closely approximated by relaxing the 'stick-where-you-hit' rule for DLA to allow particles to make a few hops around the cluster's periphery before they become immobile. But, ideally, a model of city growth should not just ape the form: it should make sense in terms of the physical processes involved. We know, for example, that urban developments don't really bounce from site to site before finally coming to rest. Batty and Longley explored a variation on the DLA model called the dielectric breakdown model (DBM), which we will encounter in the next chapter as a description of cracks and sparks. DBM clusters grow not by accumulating wandering particles at their edges but by pushing their way out from a central point. That sounds more like city growth: new development spreads outwards to colonize the surrounding land.

DBM clusters again show growth instabilities that favour extension at the tips. But their shapes can be 'tuned' from highly tenuous and almost linear (that is, almost one-dimensional) to dense and more circular (almost two-dimensional) by altering the strength of the instability—which is to say, the likelihood that new growth happens at the sharpest tips rather than elsewhere on the perimeter. Using this approach, Batty and Longley attempted to simulate the growth of the city of Cardiff (Fig. 2.22). They conducted a simulation of DBM growth constrained by the local geography: by the presence of the coastline to the south-east and the rivers Taff (to the west of the city centre) and Rhymney (to the east). The cluster was seeded from a point between these rivers, which acted as impenetrable barriers except at two points where there are bridges that the cluster could 'squeeze' across. The results of the model, for different strengths of the branching instability, show that somewhat realistic approximations to the

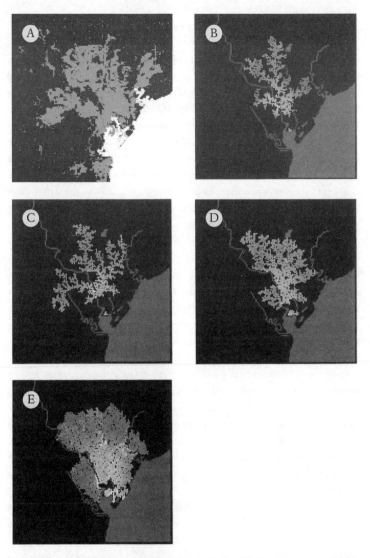

FIG. 2.22 A fractal model of the growth of the city of Cardiff (a), which is constrained by the coastline (shown here in white) and two rivers. Computer simulations for different strengths of the branching instability (measured by a model parameter η) produce urban clusters that are more or less dense (b–e). The best match with reality occurs for a value of η of around 0.75 (c). Bridges are included in the locations of the real ones, to allow the city to spill over the rivers. In these images, earlier growth is shown as lighter. (Images: Michael Batty, University College, London.)

city shape can be generated when this parameter is suitably tuned (Fig. 2.22b–e).

One of the problems with this model is that it generates only cities that form a single large fractal cluster, densest in the centre (around the central business district) and getting rapidly more tenuous as you move out. But areas of development at the edges of a big city are not always part of the main 'cluster': there are typically many little satellite towns, which may be swallowed up as the city boundaries sprawl further. You can see several clusters of this sort beyond the edges of the main cluster in the structure of London (Fig. 2.20a). Physicists Hernan Makse, Gene Stanley and Shlomo Havlin from Boston University felt that a different growth model was needed to capture this sort of structure. They allowed for the possibility of new clusters being sparked off close to but outside of the main body. These nascent centres of development do not tend to appear entirely at random, but are affected by what is happening nearby: development attracts further development. If two small population clusters grow close by one another, for instance, there is a greater than average chance that development will spring up between them: shops to serve the new inhabitants, or local businesses keen to gain a foothold in an up-and-coming area. In short, the growth of new clusters is interdependent, or what physicists call *correlated*. Makse and colleagues borrowed from physics a model that embodies such correlations, which was originally developed to describe fluid permeation in rocks. Within this model, new particles (representing units of population) are added to a growing cluster at random, as in DLA; but in addition, growth in one region enhances the prospects of growth nearby, with a probability that falls off quite sharply with distance.

This model generates a jumbled scattering of clusters of all different sizes. But real cities are firmly rooted to a core, usually the central business district. So the researchers imposed the condition that the probability of adding a new unit gets smaller the further it is from a central point. In effect this means that the model superimposes two growth processes: one that generates a single compact, roughly circular city cluster, and one in which local correlations between units can create new sub-clusters around the edges. The pattern that emerges depends on the relative strengths of these two processes, and can vary from a roughly symmetrical, circular

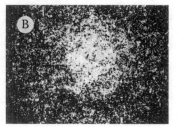

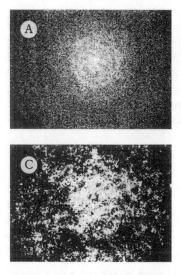

FIG. 2.23 Computer simulations of urban growth using the correlated percolation model. For increasing degrees of correlation between the growth units (top to bottom), the shape changes from more or less circularly symmetrical (*a*) to increasingly fragmented and clumpy (*b, c*). (Images: Hernán Makse, Schlumberger-Doll Research, Ridgefield, Connecticut.)

distribution of units to a clumpy form decorated with sub-clusters and tendrils (Fig. 2.23). It is clear straight away that these latter shapes, in which the correlations are strong, look more realistic than those produced by DLA or DBM, at least for a city like London. The model can also do a pretty good job of imitating how city shapes evolve over time (Fig. 2.24).

It seems, then, that randomness (uncoordinated local decisions about development, akin to the disorderly aggregation of particles in DLA) can generate something like the clumpy, messy sprawl of cities. But randomness alone is not enough to explain everything about the forms of cities. In Batty's words, it is modified by a 'deeper order', derived from growth instabilities and correlations and manifested in the power laws that define the structures. This order seems to be of precisely the same sort that we find in many natural growth patterns.

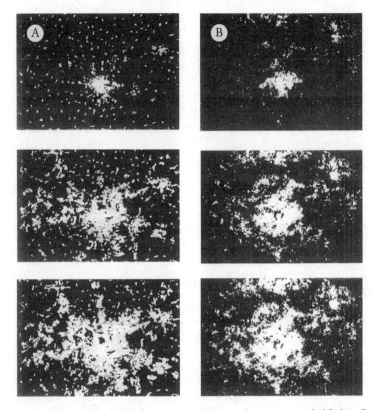

FIG. 2.24 The growth of Berlin from 1875 to 1945 (a, top to bottom) is mimicked fairly well by the growth of a city in the correlated percolation model (b). (Images: Hernán Makse, Schlumberger-Doll Research, Ridgefield, Connecticut.)

Batty suggests that rather simple concepts from DLA and DBM do a fair job of accounting for the shapes of cities that sprung up during the early industrial era, which tended to be single clusters organized around the central business district. But the 'correlated percolation' model of Makse and his co-workers is better suited to capturing the growth and form of post-industrial cities in which new communications technologies have made work in the centre of the city less prevalent and the urban landscape less centralized.

The message for planners is salutary, for it appears to question whether centralized planning for an 'organism' as complex as a city can ever succeed. In the 1960s, planners sought to influence the way in which London encroached into the surrounding countryside with a Green Belt policy that would restrict urbanization. Yet there is no sign that these policies have had any effect on the city's growth, which has gone right on expanding to the tune of the same mathematical laws. It will take more than this, it seems, to undermine the inexorable physics of cities.

JUST FOR THE CRACK

Clean Breaks and
Ragged Ruptures

Fingal's Cave on the island of Staffa, off Scotland's west coast, is the kind of place that left early nineteenth-century Romantics pondering the Sublime. You can sense as much in the characteristically storm-ridden depiction by J. M. W. Turner, and also in the luxurious harmonies of Felix Mendelssohn's tone-poem *The Hebrides*, composed after a trip to Staffa in the 1820s. This geological oddity also made an awe-inspiring impression on Joseph Banks, president of the Royal Society, when he sailed there in 1772 during an expedition to Iceland. 'Compared to this', he said,

> what are the cathedrals or palaces built by men! Mere models or playthings, as diminutive as his works will always be when com-pared with those of nature. What now is the boast of the architect! *Regularity*, the only part in which he fancied himself to exceed his mistress, Nature, is here found in her possession, and here it has been for ages undescribed.

For Banks saw that the entrance to the cave was flanked by great pillars of rock with almost perfect hexagonal cross-sections, designs of such regu-larity that they could almost have been fashioned by Platonic masons (Fig. 3.1 and Plate 4).

There is a rather more accessible example of this striking geological pattern at the Giant's Causeway in County Antrim, on the coast of

FIG. 3.1 The natural pillars of Fingal's Cave on the Scottish island of Staffa. (Photo: Lucas Goehring, University of Toronto.)

Northern Ireland (Fig. 3.2 and Plate 5). In legend, these two formations are part of the same structure: a causeway from Ireland to Scotland built by the legendary third-century Irishman Fionn MacCumhaill (Finn MacCool, or Fingal), a veritable giant and leader of the band of warriors called the Fianna who guarded the Kings of Ireland. Finn made the rocky road across the Irish Sea so that he might do battle with a rival giant of Scotland named Benandonner. When he got there, however, Finn discovered that Benandonner was a much bigger giant than he, and so he crept sheepishly back home. But Benandonner crossed over the causeway to find him, whereupon Finn posed as a baby, attended by his wife. If this is the size of Finn MacCool's baby, thought Benandoner, how big must Finn himself be? And he ran in panic back to Scotland, demolishing the entire causeway except its beginning and end.*

*Some versions of the legend find Finn's ruse too timid, and have him leap up and bite Benandonner's hand, chasing him all the way back to Scotland while hurling great lumps of earth, one of which became the Isle of Man.

FIG. 3.2 The Giant's Causeway in County Antrim, Northern Ireland (a, b). The polygonal cross-sectional pattern of the columns contains mostly six-sided shapes, although not in a perfect honeycomb (c). (Photos: Stephen Morris, University of Toronto.)

This is the way with natural patterns: we must find an explanation for them even if it need be magical and valorized by legend. We tell ourselves that these things were constructed with purpose by intelligent agents, for how could wild nature be capable of such craft?

Of course, nothing of that sort appears in D'Arcy Thompson's description of the hexagonal columns of Staffa and the Giant's Causeway. He explained that these rock formations appear in basaltic lava as it cools, contracts and cracks: 'rupture ... shatters the whole mass into prismatic fragments', he said. 'However quickly and explosively the cracks succeed one another, each relieves an existing tension, and the next crack will give relief in a different direction to the first.' This is all very fine, but Thompson does not really account for why the cracks create such a remarkable pattern: a series of vertical columns, each with a polygonal cross-section that seems most often to be hexagonal.

What these formations tell us with irrefutable force is that cracks are pattern-formers. Seldom, however, does fracture produce anything quite so regular and ordered. When we think of cracks, what comes to mind are random branching fractures that may intersect in reticulated networks (Fig. 3.3). These are familiar traceries, often to our dismay—they are the marks of old age and decay, the webs of failure and regret and disaster, all of which makes it of paramount importance to know how they arise and what guides their wandering paths. In this much, D'Arcy Thompson was able to make little progress. In view of what we have seen so far concerning branching patterns, can we now do better?

WHY THINGS BREAK

It is only recently that scientists have begun to understand why cracks form. For a long time, the science of fracture and failure of materials limped along with a ragbag of concepts that could do little to predict what was seen in the real world. This was much more than an academic embarrassment, for while scientists remained largely ignorant of what makes a material tough they had no systematic method for devising the strong and tough materials that technology demands. They would earnestly apply what seemed to be sensible criteria, only to end up with substances 'about as strong as stale hard cheese', in the words of British

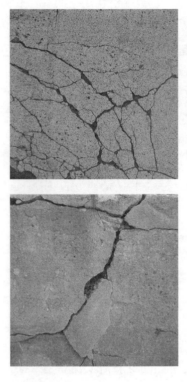

FIG. 3.3 Cracks form a variety of branching patterns.

materials scientist James Gordon. On the other hand, their experience with new materials that were genuinely strong sometimes flew so much in the face of what seemed like common sense that they had some persuading to do. Gordon recalls the response of an Air Marshall during the Second World War to the idea that Lancaster bombers were to have glass-fibre domes: 'Glass!—Glass! I won't have you putting glass on any of my bloody aeroplanes, blast you!'

We can understand the Air Marshall's feeling about glass aeroplanes, because we know how readily glass shatters. But one can offer apparently sound scientific arguments for why few materials should be better for making aircraft than glass. It consists of disordered silicon dioxide, the same stuff as sand and quartz but melted to break up the regular crystal structure and then cooled quickly so that the atoms become all but immobile before they can pack together in an orderly manner. The chemical bonds between silicon and oxygen atoms are extremely strong, not far off the strength of those between carbon atoms in diamond. You need to expend a lot of energy to pull them apart. So why isn't glass nearly as strong a diamond?

Well, the Air Marshall was wrong about glass fibres—they are very strong. But our naive chemical reasoning is wrong about glass too: it breaks rather easily. What is going on?

A stiff, brittle material like glass is tough so long as cracks cannot get started. But they can be launched from the tiniest of origins: a mere scratch may act as the seed for a flaw that shoots through the whole material. Window glass inevitably contains innumerable little scratches on its surface, any one of which can give birth to a crack that spreads with catastrophic speed and vigour. Gordon helped to show in the 1950s that only very minor rubbing of glass against another hard material is sufficient to cover the surface with microscopic cracks. The reason glass fibres are so strong is simply that they have a much smaller surface area than a plate of window glass, and so have far fewer of these minute flaws. The thinner they get, the fewer flaws there are. So cracks have nowhere to start from. In analogous fashion, Galileo (who was very interested in discovering why things break) found that the shipbuilders of Venice paid more attention to the construction of big ships than small ones—because, they told him, big ships break more easily. There are simply more places on a big ship where a crack might start.

But why, once a crack is initiated by a tiny imperfection, does it then grow with such awesome speed, if the bonds between atoms are really so strong? Knowing the energy contained in a single chemical bond in glass, it is a simple matter to calculate what the theoretical strength of glass ought to be, assuming that fracturing the glass means breaking all the bonds along the crack's path. The puzzle was that the observed strength is typically about a hundred times smaller than this calculation suggests it should be. In the 1920s the British aeronautical engineer Alan Arnold Griffith, working at the Royal Aircraft Establishment in Farnborough, had a critical insight into the problem. He was at the time laying the foundations of glass-fibre technology by drawing heated glass rods into thin threads—work that Gordon later developed at the same institution. Griffith found that the strength of very thin glass fibres is extremely high, approaching the value implied by the calculation of chemical-bond strengths. So the question was not so much why glass fibres are strong but why normal glass in bottles and windows is so weak. How can a minor scratch confined to the surface be responsible for disastrous failure?

During the previous decade, the engineer Charles Inglis had investigated why Britain's iron ships were alarmingly prone to cracking apart. He showed that if a plate of material is stressed by bending or stretching, the stresses around a hole can be much greater than those through the rest of the material: there is 'stress concentration' at the flaw. Griffith realized that the same would apply to microscopic scratches on the surface of a brittle material. He calculated that a crack just one-thousandth of a millimetre long and so narrow that it cleaves a single chemical bond at a time as it advances has a stress at the tip about 200 times that elsewhere in the material. In other words, a stress 200 times smaller than that required to break the chemical bonds in the flawless material will suffice to set bonds snapping at the crack tip. What's more, if the crack lengthens without changing its width, the stress concentration at the tip gets even greater: the growth is self-amplifying. What Griffith had demonstrated is that fracture is *nonlinear*: the effect does not follow in proportion to the cause. A tiny crack turns a modest stress into a catastrophic one.

JAGGED EDGE

We have seen already that self-amplifying instabilities tend to have other consequences too: not only does growth foster more growth, but the resulting form becomes prey to random perturbations. Small, chance irregularities generate new appurtenances: wriggles, wobbles, branches. That is just what we find in cracks. Griffith's work suggested that a long, narrow crack initiated at a notch in a brittle material will cut like a knife straight through the material when it is stressed. This, however, is the exception rather than the rule. A glazier can make a clean, straight break through a sheet of glass by first scoring a shallow scratch along the path of the intended fracture, but without this guidance the crack is more likely to be erratic: ragged or crazy-paved.

A typical brittle crack grows in three stages. During its 'birth' from an initiating notch, it accelerates in less than a millionth of a second to reach a speed almost equal to the speed of sound. During 'childhood' the crack continues to accelerate, but more slowly. All this time the crack is straight and the fractured surfaces in its wake are smooth and mirror-like, but once the crack speed rises above a certain threshold, a mid-life crisis sets in: the velocity starts to fluctuate wildly and unpredictably, and the crack tip veers from side to side. Then the fracture surfaces become rough, sprouting a forest of small side branches.

In 1951 the Cambridge metallurgist Elizabeth Yoffe uncovered a clue about why cracks may wander erratically. She found that when a crack tip moves at virtually the speed of sound, the stress field ahead of the tip starts to flatten out and develop bulges pointing in different directions from the tip's motion. These new stress concentrations could make the tip deviate from its straight path. John Willis at Cambridge showed in the 1960s that in fact the largest stresses at the tip of a fast-moving crack point at right angles to its direction of motion, suggesting that the tip should be constantly turning corners.

Yoffe's analysis implies that windows would not shatter into jagged shards, but would simply split cleanly, if the cracks did not travel so fast. But that's not necessarily so, for even cracks that grow very slowly may take complicated paths. In 1993 the Japanese researchers Akifumi Yuse and Masaki Sano at Tohoku University developed a method for growing cracks that move at just a few centimetres per second, which is much slower than those passing through a brittle material as it shatters. They sent these slow cracks through flat strips of glass by using heat to induce stress and lowering the strips slowly through a heater into a bath of cold water. As anyone knows who has mistakenly put a hot glass dish from the oven into cold washing-up water, this abrupt cooling can shatter glass. When hot, the material expands; when cooled, it shrinks. At the boundary of expanded and shrunken material there are large stresses which can cause cracking. But the cracks advance only where there is a sharp change in temperature over a small distance. By varying the rate at which they lowered the glass strips, the Japanese researchers were able to control precisely the speed of a crack initiated from a notch at the bottom.

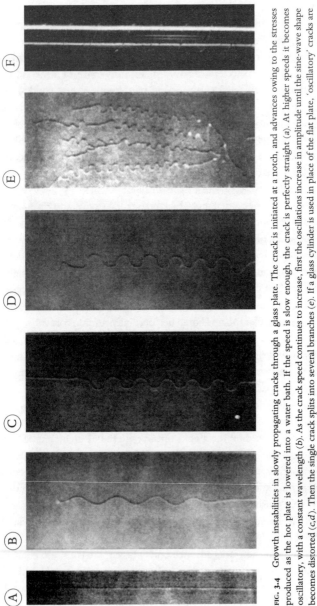

FIG. 3.4   Growth instabilities in slowly propagating cracks through a glass plate. The crack is initiated at a notch, and advances owing to the stresses produced as the hot plate is lowered into a water bath. If the speed is slow enough, the crack is perfectly straight (*a*). At higher speeds it becomes oscillatory, with a constant wavelength (*b*). As the crack speed continues to increase, first the oscillations increase in amplitude until the sine-wave shape becomes distorted (*c,d*). Then the single crack splits into several branches (*e*). If a glass cylinder is used in place of the flat plate, 'oscillatory' cracks are not wavy but instead thread around the cylinder in a helix (*f*). (Photos: Akifumi Yuse and Masaki Sano, Tohoku University.)

They found that for very slow speeds (about a millimetre per second) the cracks were generally perfectly straight (Fig. 3.4a). But if either the speed or the temperature drop between heater and water bath exceeded particular thresholds, the crack became unstable and began to wiggle—not at random, but in a steady oscillation with a well-defined wavelength (Fig. 3.4b). In other words, you can give a glass sheet a rather beautiful, undulating edge just by cracking it with heat and cold.

Yuse and Sano found that the wider the glass strip, the longer the crack's wavelength. If the strip was infinitely wide, the wavelength would be infinite too, meaning that the crack would head off at a fixed angle to the vertical and never look back. The researchers couldn't lay their hands on an infinitely wide glass strip, but they could find one without edges: a glass tube. And here indeed the crack simply travelled around the tube's axis at a fixed angle, cutting out a perfect helix (Fig. 3.4f). Such helical cracks are rumoured (this is not easy to check) to be found in frozen natural gas pipelines in Alaska, sometimes winding their way around the pipes for miles.

As these wavy cracks get faster, the oscillations become more pronounced, until finally they start to become distorted and kinked (Fig. 3.4c, d). And if the temperature drop is large enough, the wavy cracks grow branches, apparently two at a time: the crack repeatedly *bifurcates*. At this stage, the pattern starts to become more disorderly—more like the classic picture of a crack—although the regular waves can still be seen.

Thin sheets of material often break in wavy patterns, although they are not usually as regular as this. One of the most familiar is the jagged tear made by opening a letter with your finger (Fig. 3.5). Animangsu Ghatak

FIG. 3.5 Opening a sealed envelope with your finger tends to generate an oscillatory tear with ragged, sawtooth edges.

and Lakshminarayanan Mahadevan of the University of Cambridge have constructed a physicists' version of letter-opening, which reveals where this waviness comes from. They used rigid rods, ranging in width from a few millimetres to several centimetres, to rip open stiff plastic film similar to that used as transparent wrappers in packaging. The researchers pulled the rods through the sheet at a constant speed of up to 2.5 cm per second. For narrow rods, the tears were straight and smooth, rather like those made by a paperknife. But fatter rods produced evenly spaced serrations with sawtooth teeth, which have a mathematical shape known as a cycloid (Fig. 3.6a).

Ghatak and Mahadevan deduced that this shape stems from a mixture of stretching and bending as the sheet is ripped. The convex slope of a sawtooth forms as the sheet bends back against the rod in a kind of curling tongue (Fig. 3.6b). This 'bending rip' requires little energy, and it sends the

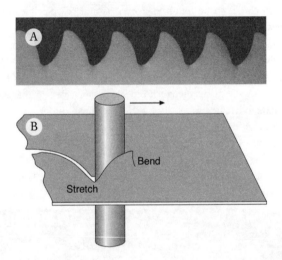

FIG. 3.6   A carefully controlled 'letter-opening' experiment using a cylindrical rod to tear a flat sheet produces a remarkably regular sawtooth tear with 'teeth' that have a shape described mathematically as a cycloid (a). This can be explained by considering how the tear bends and curls away from the advancing rod edge, until there is not enough energy to sustain this bending (at the tips of the cycloids) and the sheet begins to rupture instead by stretching (b). (Photo: a, L. Mahadevan, University of Cambridge; from Ghatak and Mahadevan, 2005.)

crack heading off at an increasingly oblique angle to the direction the rod is travelling in. But the further it goes, the smaller the region over which bending occurs, and eventually there is just not enough 'bend' left. At that point the rod, still advancing and pushing against the torn edge of the sheet, begins to stretch it, and the sheet ruptures by being pulled apart rather than bent—which means that the rip moves forwards in the same direction as the rod. Very quickly this 'stretching rip' runs ahead of the rod and so loses impetus. Then the rod catches up and begins a new bending rip, which heads away in the opposite direction from before. So each crest of the cycloid, where the rip changes direction, corresponds to the switch from bending to stretching of the sheet. The crack swings constantly from side to side, at the same time surging ahead and then slowing down like the juddering stick-and-slip of a heavy object being pushed across a floor.

Similar ruptures occur on a much larger scale when polar sea-ice sheets are pushed past grounded icebergs by ocean currents or winds. But apparent fractures in ice sheets can take an alternative path: occasionally they are found to run back and forth in rectangular crenellations that interleave like a zipper, a phenomenon called finger rafting (Fig. 3.7a). These patterns have a rather different origin. They form when two thin ice sheets, no more than about 10 cm thick, are pushed against one another. If the ice sheets have ragged edges, then some of these

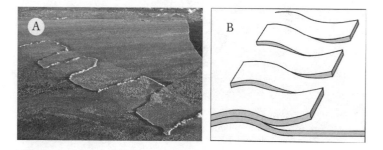

FIG. 3.7 Cracks in floating ice sheets can develop a zipper-like shape, a phenomenon called finger rafting (a). This is caused by the collision of two ice sheets, in which a protrusion on one edge that rides up over the other creates a wavy deformation to either side, generating 'zipper teeth' at regular intervals (b). (Photo: a, John Wettlaufer, Yale University, New Haven; from Wettlaufer and Vella, 2007.)

protuberances on one sheet will ride up over the other, pressing down on it at the point of overlap. The effect of such an over-riding lip is to create small corrugations on either side in both sheets, running perpendicular to the boundary between them. The troughs of these ripples in one sheet coincide with the crests in the other, so that a single lip induces other overlaps regularly spaced to either side. These split into the 'fingers' of the zipper (Fig. 3.7b). John Wettlaufer of Yale University and Dominic Vella of Cambridge University explained all this in 2007 and showed that the same effect can be seen in thin layers of wax pushed together on the surface of water. It is possible that similar crenellated structures form on immense scales along geological faults when two tectonic plates converge.

A MATTER OF CHANCE

These are strange cracks, and very different from the jumble of spidery lines we tend to associate with fracture (Fig. 3.3). The striking thing about

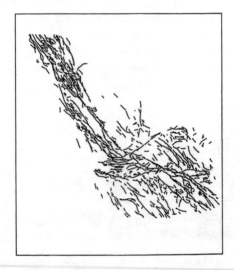

FIG. 3.8 Fractal cracks occur on many scales. Here is the network of fault lines that surrounds the San Andreas fault—the image encompasses many kilometres, although you could probably just as easily imagine it to be a diagram of cracks in, say, an old layer of paint on a window frame.

those patterns is that they cover a huge range of size scales: some are visible only under the microscope, some only from satellite imagery of geologically vast terrains (Fig. 3.8). And many crack patterns look much the same over a wide range of scales: as we zoom in at increasingly higher magnification, they don't appear to change very much. We merely see ever finer details that we could not make out before. What this implies is that fracture patterns are scale-invariant fractals.

These patterns, unlike the regular wavy cracks described above, involve a strong dose of randomness. Some of this may stem from the nature of the cracked materials themselves: rocks are typically haphazard compactions of grains of many different sizes and shapes, welded together at their boundaries; cement and porous rocks like sandstone are shot through with random networks of pores; hard, brittle plastics contain a chaotic tangle of polymer chains. But some randomness seems intrinsic to the way a crack moves: as we saw, the path of a crack tip is potentially unstable, weaving this way and that or growing branches at the slightest opportunity. In this way, even materials with perfectly ordered atomic-scale structures (crystals) can fragment in erratic, jagged shapes.

A popular model of fracture devised in the 1980s suggests how fractal branching cracks might thread their ways through an orderly lattice of particles. The model was intended by its creators, Lutz Niemeyer, Hans Jurg Wiesmann, and Luciano Pietronero at the Brown Boveri Research Centre in Baden, Switzerland, to describe a rather specific kind of 'failure': not fracture in the normal sense, but the passage of a spark through a material. In electrical devices such as capacitors, an electrical voltage is applied between metal plates or electrodes separated by a layer of insulating material called a dielectric. If the voltage is too big, a spark discharge crackles between the electrodes. This is called dielectric breakdown (Fig. 3.9a), and it usually burns out the device. In some cases it is accompanied by real fracture: the material is shattered by the flow of charge. In transparent materials the fracture pattern can leave visible tracks: a frozen snapshot of the spark's passage, which is in itself a thing of considerable beauty (Fig. 3.9b). The structure has a branched, lightning-like appearance, and indeed atmospheric lightning (Fig. 3.9c) is itself a closely related phenomenon: air acts as an electrical insulator between a charged cloud and the ground.

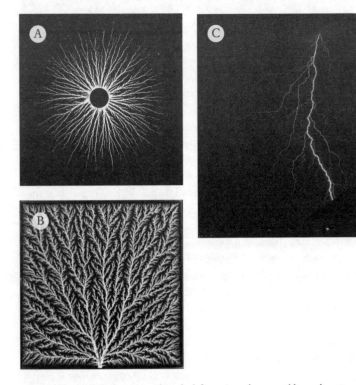

FIG. 3.9 Electrical discharges are branched formations that resemble crack patterns, as shown here in the spark pattern from an electrode on the surface of a glass plate (a). These so-called dielectric breakdown patterns can be 'frozen' and preserved when the passage of current heats up a solid substance and cracks or vaporizes it (b). The beautiful structures that result are called Lichtenberg figures. A lightning discharge in the atmosphere is also an example of dielectric breakdown, with a similarly branched pattern (c). (Photos and images: a, After Niemeyer et al., 1986; b, Kenneth Brecher, Boston University; c, Michael Mortenson.)

Electrical discharge patterns like these were studied in the eighteenth century by the German scientist and writer Georg Christoph Lichtenberg, and they are commonly called Lichtenberg figures. Lichtenberg, working at Göttingen University, was investigating the new science of electricity by building up intense charges of static and letting it discharge through a block of resin. He found that dust was attracted to the flow of charge, accumulating on the resin in a way that revealed the many-fingered routes it took. Lichtenberg's experiments led him to invent an early form of electrostatic printing, based on the way powder arranges itself on a plate of material with an uneven distribution of electrical charge. He was also the first person to show conclusively that lightning is an electrical phenomenon by carrying out a hazardous version of the kite-flying experiment popularly (and probably apocryphally) attributed to Benjamin Franklin. He constructed a lightning conductor at Göttingen, and corresponded with Alessandro Volta, the inventor of the battery. (Lichtenberg noted that Volta had 'an expert knowledge of the electricity in girls.'*) But it is for his work as an essayist and aphorist that Lichtenberg is best remembered—as with many Enlightenment intellectuals, scientific research could not contain his diverse energies. When Lichtenberg said of his 'frozen lightning' that it 'transfer[s] terror into enchantment', it is a reminder of how undervalued today is the scientist who, with a *mot juste*, can distil wonder from dry experimentation. And when he compared his electrical figures with dendritic ice-fingers on a window pane in winter, we can now see that his penchant for poetic metaphor put him unwittingly on the right track.

Lutz Niemeyer and his colleagues chose to model the dielectric breakdown process by considering a regular checkerboard lattice on which charge could flow from point to point in straight lines. The discharge advances one lattice site at a time, and there are generally several directions it could travel next (Fig. 3.10). Which way does it go? The researchers assumed that the discharge passes at random to any of the next accessible points at each time step, but with a bias that depends on the strength of the electric field at that point; in other words, there might be (say) a 70 per cent chance that it will move to an adjacent site where the electric field is high, and a 30 per cent chance that it will move to a position where the field is lower. This is a reasonable thing to assume, since the passage of the spark can indeed be expected to be governed by the electric field it encounters en route. Here again we see a delicate balance of chance

---

*There can be little doubt that Lichtenberg and Volta had a good time together. 'What is the simplest method to produce a good vacuum in a wineglass without using an air pump?', Lichtenberg asked his friend at one point. 'Just pour in wine! And what is then the best method to allow the air to come back? Just drink the wine!' He added that 'This experiment will seldom fail!', suggesting that he had been careful to accumulate good statistics.

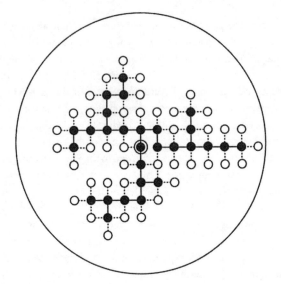

FIG. 3.10 In the dielectric breakdown model, an electrical discharge advances between adjacent points on a regular lattice. There are in general many different points to which the discharge could flow next—here I show the course of the discharge in black, and the choices for its next step in white. (The discharge began at the centre of the image, the circled black grid point.) The next step for the discharge is selected at random but with a probability that is biased by the strength of the electric field at each of the candidate points. (After Niemeyer *et al.*, 1984.)

and necessity: in the evolution of the discharge route, nothing is certain but some things are more likely than others. There is randomness, but not that alone.

The electric field around the tips of a branching discharge is higher than that in the valleys and clefts, so advance from the tips is more likely than advance from the interior of the 'spark'. That is just how it is for DLA clusters, too, and for an entirely analogous reason: in that case, the probability of a new particle striking the tips is higher. It is no surprise, then, that the dielectric breakdown model produces ramified, tenuous discharge patterns that look very much like DLA clusters (Fig. 3.11a). If the probability of the discharge flowing to a new site is assumed to vary in

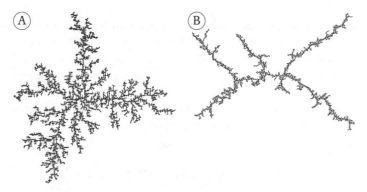

FIG. 3.11   The dielectric breakdown model generates branching fractal patterns very similar to those seen in diffusion-limited aggregation (a). They have a fractal dimension of about 1.75. This model can be adapted to describe the propagation of cracks in a brittle material, represented as a lattice of particles linked by bonds. If the bonds are considered to be a little elastic, able to stretch and relax in response to stresses, the branching pattern is more tenuous (b). (Images: a, Luciano Pietronero, University of Rome 'La Sapienza'; b, Paul Meakin, University of Oslo.)

direct proportion to the electric field at that site, then the patterns predicted by this model have a fractal dimension of 1.75—almost the same as that of DLA.

The dielectric breakdown model can be imported wholesale into a theory of fracture in disordered materials. All this entails is to regard the discharge as a crack, and the lattice as a network of interconnected atoms or particles joined by chemical bonds. At each time step, the crack takes a pace forward as a new bond breaks, and the probability of that happening at any site adjacent to the crack's periphery depends on how big the stress is there. The stress field varies in just the same way as an electric field: it is greatest at the tips of the crack and is smaller in the crevices, just as Griffith said.

But this is a very simplistic picture of fracture. For one thing, it insists that a bond must break with every time step—the only question is which it will be. In reality, though, there is no reason why that should happen if the stress simply isn't large enough. A better model could allow bonds to stretch a little without breaking, so that they are not like rigid rods but more like springs. In that case, each time a bond breaks it will release local stress as the surrounding bonds relax. 'Elastic' models like this generate a range of different fracture patterns, depending on what is assumed about the bonds' elasticity. The example in Fig. 3.11*b* shows a crack that has a much less dense network of branches than those generated by the rigid-bond dielectric-breakdown model, and looks rather more like the kind of pattern you might finds creeping ominously across the ceiling. The fractal dimension here is 1.16, confirming that the crack is less like a two-dimensional cluster and more like a wiggly line.

PATTERNS IN THE DRY SEASON

All these cracks start at a single point; but that is not always the way it works. In the hard mud of a dried-up pond, there is no centre to the network of cracks: the rupturing happens everywhere at once, creating a

FIG. 3.12   When a thin layer of material is stressed as it shrinks, it may fragment into a series of islands. This process is familiar in the mud on the bed of dried-up ponds and lakes. (Photo: Sean McGee.)

reticulated web (Fig. 3.12). As the wet mud of the pond bed becomes exposed and dried, water withdraws from between the silt particles and they draw closer together. The whole surface layer contracts. But it cannot simply shrink like a dried-up piece of fruit, because this dry layer adheres to the damper mud below, which is still expanded by water. This means that stresses build up everywhere in the surface layer. When the stresses get big enough, cracks start to appear, creeping through the hard mud and intersecting to carve it up into islands.

This kind of cracking due to uniform shrinkage (or expansion) of a thin layer of material 'pinned' by adhesion to another surface is very common. It is apt to afflict a coat of paint as the material on which it sits expands or contracts: a piece of wood swells or shrinks as the humidity changes, say, or metal expands as it warms. This is the cause of the mesh of hairline cracks that weaves across old paintings, known to art historians as craquelure. The resulting patterns offer a subtle fingerprint of authenticity: they may vary, for example, according to where and when the picture was painted—there are French, Italian, Dutch and Flemish 'styles' of craquelure. The cracking pattern provides clues about the artist's materials and techniques, or the history of the painting: how it was handled, transported, and changes in its ambient environment. So art conservators have developed sophisticated methods for digitally scanning paintings to identify and classify the craquelure pattern. Forgers, meanwhile, have attempted to mimic it by baking their works; they have even been known to resort to the crude method of painting a fine web of craquelure by hand—a deception that might fool the eye of a careless buyer but which would be immediately obvious under a magnifying glass.

Cracking of thin layers poses a big challenge for several advanced technologies. Surface coatings that are deposited 'wet' to protect or modify a material, for example conferring wear-resistance or low reflectivity, may shrink as they dry or cool, threatening to leave them cracked and flaking. Integrated microelectronic devices often incorporate a thin film of one crystalline material (an insulator perhaps) laid down on top of another (a semiconductor, say) in which the spacing between the atoms is slightly different, forcing the top layer to expand or contract. So there are good reasons for wanting to understand what determines the crack patterns that form.

Arne Skjeltorp from the Institute for Energy Technology in Norway has explored this type of fracture with what one might call an 'ideal mud': a suspension of microscopic spheres of polystyrene in water. All the micro-spheres are the same size, each measuring just a few thousandths of a millimetre in diameter, and Skjeltorp trapped a single layer of them between two sheets of glass and let the water slowly evaporate. The particles clump together just like silt particles in pond mud, and the layer contracts as water disappears. But the identical size and shape of the spheres means that they pack together more regularly than silt,

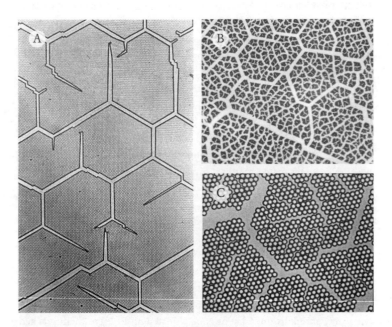

FIG. 3.13   The cracks in a layer of microscopic plastic particles suspended in water—a kind of 'ideal mud'—as it dries. Because the particles are identical in size and packed in a hexagonal array, the cracks tend to follow the lines between rows of particles and therefore intersect at angles of about 120°. This is particularly evident in the early stages of cracking (a). The final crack pattern (b, c) looks similar at different scales of magnification, at least until we reach the scale at which individual particles can be seen (c). The region in b is about one millimetre across; that in c is ten times smaller. (Images: Arne Skjeltorp, Institute for Energy Technology, Kjeller.)

forming orderly hexagonal arrays that are also good mimics of atoms packed in crystalline films.

Skjeltorp found that as the layer dries, it fractures into complex 'crazy paving' patterns like those of dry mud (Fig. 3.13). The web of cracks looks similar at different scales of magnification: again it is fractal, with a fractal dimension of about 1.68. But there are signs of orderliness here: the cracks have preferred directions, tending to intersect at angles of 120°, particularly in the early stages of drying. This simply reflects the symmetry of the underlying lattice of particles: the cracks tend to open up between rows of particles. The particles in mud are packed together in a much more disorderly fashion, and so the shapes of the final islands are less regular. The physicist Paul Meakin at the University of Oslo has adapted the 'elastic' dielectric breakdown model to this situation, and found that it produces crack patterns similar to those observed (Fig. 3.14).

But many more familiar examples of cracking are rather different from this. For mud drying on a lake bed, paint and varnish drying on canvas or

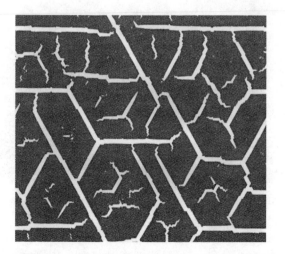

FIG. 3.14 A modified form of the dielectric breakdown model can reproduce the fracture patterns seen in contracting thin films. (Image: Paul Meakin, University of Oslo.)

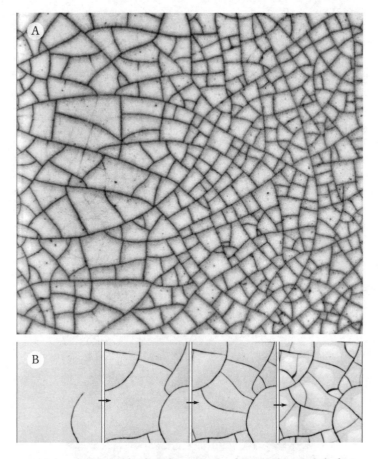

FIG. 3.15 A crack pattern in the glaze of a ceramic plate forms a web in which the fracture lines tend to intersect at right angles and the islands have an average of four sides (a). The cracking happens in a hierarchical manner: first, the longest cracks form, and gradually the spaces between them become laced with bridging cracks (b). (Photos: Steffen Bohn, The Rockefeller University, New York.)

wood, or a ceramic glaze hardening and ageing on a piece of pottery, the cracking layer is fixed to a surface on one side while being freely exposed to air on the other. In such cases, the crack pattern often takes on a rather different appearance (Fig. 3.15a). Some cracks seem to advance in straight or curved lines without splitting or bending sharply for long distances, while other cracks divide up the space in between. This breaks the material into fragments that typically have four sides, many of which are square or rectangular. In the final patterns these domains have a 'typical' size that depends on the thickness of the cracked layer—which means that this crack pattern is *not* a fractal. And despite the right-angled nature of the junctions, the pattern is not simply a square or rectangular grid, like the grid-iron street plan of Manhattan: the junctions are staggered, so that domains tend to share edges not with four but with six neighbours.

Steffen Bohn of the Rockefeller University in New York and his co-workers have shown that in this case the crack network forms in stages: first by the appearance of long, smooth cracks, and then by the successive division of the space in between them as smaller cracks bridge the gaps (Fig. 3.15b). When an advancing crack intersects with an existing fracture line, stress is relieved most effectively if the junction makes a right angle. The new crack may bend in order to satisfy this condition. Bohn and his colleagues have shown how this pattern of cracks can be analysed to deduce the order in which the cracks formed, and thus to reconstruct the history of the fracturing.

The process is not unlike the way new urban roads are built between existing ones, which is why the resulting crack pattern shows similarities with town road networks. This is seen most clearly in older cities where the streets were laid down spontaneously, without a planner's overarching vision: in Paris, say, but not in New York (Fig. 3.16). Here the initial long cracks are the oldest roads, leading from the city centre to surrounding villages or abbeys. Paths and lanes sprang off from these main arteries, perhaps initially to give carts access to fields. These informal routes became streets as the urban centre expanded. Right-angled T-junctions offer the most direct routes between one artery and another,

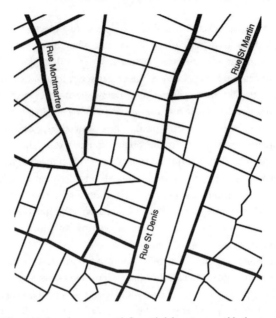

FIG. 3.16 Hierarchical crack patterns with four-sided domains resemble the street networks seen in old cities that have grown spontaneously, as shown here for a map of Paris from 1760. (From an image kindly supplied by Steffen Bohn.)

and so the road network evolves in a hierarchy of subdivisions, just like a cracking film. If Bohn's theory is right, then, a street map too encodes its own history.

THE DEVIL'S HONEYCOMB

In the polygonal islands of these crack patterns, traced out by cracks that intersect with near-geometric precision, we might begin to discern the outlines of the prismatic columns of the Giant's Causeway. But the answer is not as simple as that. For one thing, those rock pillars did not form in a thin layer, but are the result of a hexagonal web of fractures that passes downwards almost unchanged through many metres of basalt. And we cannot ascribe these hexagons to any underlying hexagonal packing of constituent particles, as in the case of the drying layers of identical micro-spheres. Any explanation will have to work a little harder than that.

Whatever the cause of the causeway, it must be a rather unusual one. Basaltic lava has flowed from many places on the Earth, but few of these igneous deposits have cracked in such regular array as they cool (another example is the Devil's Postpile in California). Yet the process has been seen to occur in other materials too. In 1922 a British optical engineer named J. W. French studied the cracking of starch—a slurry of tiny particles, not dissimilar to mud—as it dries. His experiment was part of a programme designed to mimic the cracking of glass, but French found that drying starch breaks up into columns: a kind of mini-Giant's Causeway, as he recognized. The work was all but forgotten until Gerhard Müller of the J. W. Goethe University in Frankfurt repeated the experiment in 1998. He found that layers of starch several centimetres thick fractured into columns a few millimetres wide.* Stephen Morris and his colleague Lucas Goehring at the University of Toronto in Canada repeated this experiment in 2006, with the same result (Fig. 3.17).

The cracking of the Giant's Causeway began at the top, where the heat escaped and the molten rock was therefore coolest. The network of cracks formed there will have gradually advanced downwards into the solidifying rock bed. But this does not seem to have happened smoothly: instead, the fracture pattern progressed downwards in layered steps, each successive layer freezing and cracking before the next. We can see that this was so because it has left horizontal striations in the surfaces of the basalt pillars.

This gave physicists Eduardo Jagla of the Centro Atómico Bariloche in Argentina and Alberto Rojo of the University of Michigan a vital clue to the way the prismatic columns were created. In 2002 they proposed that the crack network in the initial layers was much more irregular, but that it became progressively more orderly and polygonal as it propagated deeper. For a fixed total length of cracks, a polygonal, roughly hexagonal network is more effective than a random jumble at releasing the stress that builds up in the cooling and contracting layer of rock. This is somewhat comparable to the way that a hexagonal honeycomb network of soap films in a

---

*I know of no other paper with an acknowledgement like this one: 'I thank my wife for her patience with my kitchen interferences.'

FIG. 3.17 An artificial Giants's Causeway made from columns of corn starch left to desiccate in a thick layer (a). The ruler at the top, marked in millimetres, shows the scale. A columnar structure like this in basaltic rock can be found in other parts of the world besides Staffa and the Giant's Causeway, such as in this formation near Banks Lake in Washington State (b). (Photos: Stephen Morris, University of Toronto. b, from Goehring et al., 2006.)

foam provides the least total surface area and therefore the lowest-energy configuration, as I explained in Book I. The uppermost layers of the rock cannot 'find' this optimal pattern, however, because the cracks just start anywhere at random and advance until they meet another. But as each successive layer freezes and ruptures, this crack network gets reorganized so that it takes on an increasingly polygonal shape. Once such a network is attained—and it does not have to be perfectly hexagonal, but only roughly so—it remains fixed, since further reorganization would not significantly improve the way stress is relieved. At that point, then, the network stays identical from one layer to the next, creating vertical-sided columns.

A process like this in which irregular cracking of the uppermost layers becomes marshalled into a stable polygonal network as the cracks descend was in fact suggested in 1983 by Irish physicists Denis Weaire and Conor O'Carroll in Dublin, who in turn got the idea from the American metallurgist Cyril Stanley Smith. (Weaire says that the notion of descending 'crack layers' goes back further still.) But they had no evidence that it could work. Jagla and Rojo showed that it might: they tested their idea using computer simulations of a layer-wise cracking process. The first layer had a jumbled web of branching cracks, but by the eighth layer this had become tidied up into polygons that then changed hardly at all in lower layers (Fig. 3.18). Most of these polygons are six-sided (the number ranges from four to eight), and they all have roughly the same size. This looks very much like the pattern seen in the Giant's Causeway (Fig. 3.2c). Better still, the researchers showed that both the relative proportions of polygonal cross-sections with different numbers of sides, and the relationship between the number of sides and the area of the polygons, closely match those observed in the geological formation.

If this picture is correct, where are the original, more disorderly upper layers of the Giant's Causeway? As with many rock formations, they have most probably been stripped away long ago by wind, rain, and sea. But Morris and Goehring verified that prismatic pillars may appear in this manner with their experiments on desiccating starch, where they found that an early jumble of descending cracks does indeed find its own pseudo-geometric order (Fig. 3.17a). The final stable pattern depends on several factors, in particular the cooling or drying rate of the material. Morris and Goehring found that each set of experimental conditions seems to offer a

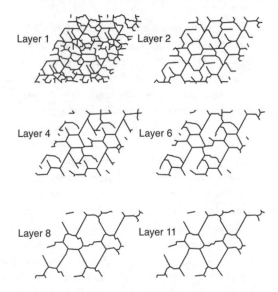

FIG. 3.18 Computer simulations of the successive ordering of cracks in a substance that solidifies layer by layer show that they can evolve from a rather random array to one that has largely hexagonal domains. (After Jagla and Rojo, 2002.)

range of different average pillar widths, and it is not obvious which of these will be selected in a particular case: this may depend, for example, on the previous history of the process. The sizes of all the columns can change suddenly as some descending cracks run out of steam and the others rearrange themselves. In other words, specific polygonal crack patterns may be *selected* from a range of possibilities: all may be more or less hexagonal, but the scale varies. You could say that the propensity for patterning is inherent in the process of solidification, but the precise pattern that results (and in particular, the size of its elements) depends partly on the whims of the elements. So once again, chance and necessity supply the choreography.

WATER WAYS

Labyrinths in the Landscape

R ivers, like all things linked to water, are popular metaphors. They speak of time and life and journeys, of blood and tranquillity and turmoil. The metaphors shift, however, and are liable to trip us up. If life is a river, what is the Styx? Rivers nurtured the earliest civilizations, but periodically decimated them.

When the biologist Richard Dawkins compared evolution to a river in his book *River Out of Eden*, he had in mind both the notion of time's flow and the luxuriant branches of the phylogenetic tree that connects all species. This is a vivid image, but it's best not to think too hard about it—for the river's source lies in its tips, not in the channel to which all tributaries converge.*

That is the odd thing about rivers: their networks grow in the opposite direction to the way the water flows, their headwaters cutting backwards into rock. There is a very real sense in which we can regard a river as a crack, propagating slowly across a range of hills or mountains.** Yet there

---

*Moreover, rivers flow in a particular direction, whereas some biologists, such as Stephen Jay Gould, argue vigorously that there is no direction in evolution: it branches, but is 'going' nowhere.
**In the real world it's actually a little more complex than this, since rivers do not just grow from the tips. For example, sometimes tributaries of one channel can become captured by another, causing them to reverse the direction of their flow.

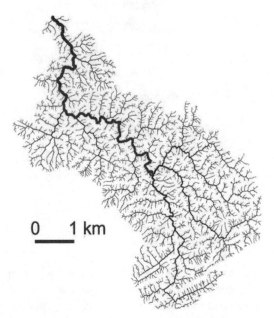

FIG. 4.1 River networks: geomorphological cracks on a grand scale? Here is the drainage basin of the Dry Tug Fork in California. (Image: Andrea Rinaldo.)

is no 'pressure' pushing the tips forward. All the same, the result (Fig. 4.1) is a pattern that looks rather like the branched formations we have seen already: not only cracks but sooty aggregates, bacterial colonies, and electrical discharges.

In fact, rivers are arguably the grandfather of branching patterns: the first that were ever contemplated in terms of their formal shape. Leonardo da Vinci, whom we encountered in Book II as a pioneer of fluid flow patterns, sketched the most extraordinary topographical maps of rivers and their watersheds, in which the mountains are represented not, in the medieval manner, as so many stylized conical profiles, but with shading that shows something like a contour or tree line (Fig. 4.2). It is as if he really took to the skies in one of his flying machines to obtain this bird's eye view, from where the ferny fractal fronds of the river basin look eerily like those evident in today's satellite imagery (Plate 6). Leonardo had good

FIG. 4.2 Leonardo da Vinci's 'aerial' sketch of a river network and the surrounding topography looks strikingly like the fractal forms seen in today's satellite imagery (*Plate 6*).

reason to strive for this accuracy, since these were most probably plans for his hydraulic engineering schemes, such as the construction of canals on the River Arno in Tuscany. Yet for all his practicality, Leonardo's vision was also informed by his metaphysical convictions: when he called rivers the 'blood of the earth', this allusion to the venous structure of the network was not just pretty word-play but was rooted in Neoplatonism. He firmly believed that structures found on a grand scale in nature—the macrocosm—would be reiterated in the body of man, the microcosm. We will see shortly that he was not mistaken in this regard. Not only do these forms look alike, but they may well have the same ultimate cause.

SCALING UP STREAMS

For geomorphologists (those scientists who study the shapes of landforms) the branching patterns of rivers were one of the most prominent features of natural landscapes, and demanded explanation. The first attempt to formulate one was made by the American hydrologist and engineer Robert E. Horton in the 1930s, who proposed that there are universal 'laws of drainage network composition'. These laws are now generally cited in the modified form defined in 1952 by geomorphologist Arthur Strahler. Each branch of the river network is assigning an 'order' that signifies its position in the hierarchy. The streams at the tips, which themselves have no tributaries, are first order. Where two first-order streams join, the resulting stream is second order; and in general, the meeting of two streams of a given order signals the beginning of a stream of next highest order (Fig. 4.3). If a lower-order stream flows into a higher-order stream, the former terminates but the latter's order is unchanged.

Horton claimed that these stream orders obeyed certain mathematical regularities. His 'law of stream numbers' states how the number of streams of a particular order depends in a predictable way on the order. It is clear from looking at a drainage network that there are fewer higher-order streams than lower-order. Horton expressed this with mathematical precision: the number of streams of order n is proportional to the inverse of a constant C raised to the power n. The number of second-order streams is proportional to $1/C^2$, third-order to $1/C^3$, and so on. This law is an example of a power law or scaling law (see page 35). Another way of

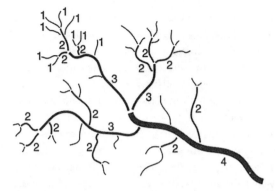

FIG. 4.3 The hierarchy of river network elements in Strahler's modification of Horton's classification scheme. The order assigned to each branch increases as one progresses from initial tributaries to the main river channel.

expressing this idea is to say that the number of streams in each order is a constant times the number in the next highest order. The number of first-order streams in a particular network might, for example, be four times the number of second-order streams, which is itself four times the number of third-order, and so on. This would mean that each stream has, on average, four branching streams of the preceding order.

Horton also proposed a scaling law for stream lengths: the average length of a stream of order n is proportional to a (different) constant D raised to the power n. (Or again: the average length for each order is a constant times the average length of the preceding order.) Thus, streams of higher order are longer—again what you would anticipate intuitively from looking at the drainage map. A third scaling law relates the down-stream slope of a stream to its order. In 1956 the American geomorphologist Stanley Schumm proposed a fourth law, in the same spirit: the area of the drainage basin feeding a stream with water increases with stream order in the same way as stream length: that is, proportional to a constant raised to the power n. And a year later a geologist named John Hack added one more scaling relationship, pointing out that the area of the full drainage basin for a network increases in proportion to the length of the principal river (that is, the highest-order element of the network) raised to the power of about 0.6. Hack's law seems to be more or less true for drainage networks ranging in size from those produced in small laboratory experiments to those almost as big as the Amazon. But there is some dispute about the precise value of Hack's exponent: some estimates place it closer to 0.5, while others think it does not really have a universal value at all, but varies slightly from place to place. It appears, in fact, that the spread in observed values of Hack's law is related to the fact that river basins seem to become elongated (they get narrower) as they get bigger.

These scaling laws are really expressions of the fractal, self-similar character of drainage networks. They are reflections of the fact that the structures look the same over a wide range of magnification scales, so that you might not know from an aerial photo if you are looking at an area a mile across or a hundred miles. Where does this fractality come from? And can we explain why the networks follow these particular scaling laws, and not others?

When Horton first reported his laws they were regarded with awe, as though uncovering a profound natural regularity. But in 1962 Luna Leopold and Walter Langbein showed that randomness alone is enough to ensure that these relationships hold for *any* branching network. They proposed a model of stream formation based on that developed by Horton himself, according to which networks emerge as rain falls onto a gently undulating surface. Wherever rain delivers more water than can be removed by filtering down through the rock bed, water accumulates on the surface and flows down the steepest gradient. These are not really streams, but so-called 'rills'—they don't 'go' anywhere in particular, but just cover the land surface with little gullies that grow bigger and deeper as the water erodes the surface. As rills grow larger, they begin to merge. In the model developed by Leopold and Langbein, rills form at random and larger channels arise from the merging of smaller ones. Because the surface topography is random, the rills evolve in random wiggles, constrained only by the fact that, like streams, they cannot re-cross their own tracks (a property called self-avoidance). This model generates networks that obey Horton's laws as if by magic, even though its ingredients reflect only the barest details of the real geological processes.

In 1966 the geomorphologist Ronald Shreve showed that in fact Horton's laws are extremely likely to result from any process that connects at random a given number of stream sources in a drainage basin into a network. And the geomorphologist James Kirchner demonstrated in 1993 that even randomness is not essential: almost *every* kind of branched network conceivable obeys Horton's laws, not just those arising from random processes. In other words, Horton's laws don't really tell us anything at all about the fundamental patterns of stream networks. They are probably instead an inevitable consequence of the scheme that Horton (and subsequently Strahler) used to break down the networks into fundamental units of different order. So it is no good testing a particular model of drainage network development by seeing if it generates Horton's laws, because just about any model will do that.

This is a salutary tale. You see a pattern and you discover that it conforms to a particular mathematical description, and so you think that the maths captures the essence of the pattern. But it may be that lots of other patterns follow that same mathematical law, while differing in other ways. That is a particular hazard of research on fractals: a fractal dimension is a useful measure, for example, but it is not a unique fingerprint of the pattern.

INVASION OF THE HIGHLANDS

In any case, it is now clear that drainage networks do not usually form by the random appearance of rills followed by their merging. Instead, a network grows from the heads (tips) of the channels, where erosional processes cut back into the rock. To understand why networks have the form they do, we must therefore focus on what is happening here at the stream heads.

Cutting into rock requires energy. In stream networks this comes from the kinetic energy of rainwater or meltwater flowing downhill. The energy input is greatest where the water flows fastest and most abundantly: where steep slopes converge in the funnel-shaped head of the channels. Downstream, the flow is more sluggish and erosion is less urgent. So the network grows from the branch tips—just as it does for cracks, lightning, or viscous fingering, where the branch tips are the places of steepest gradient in stress, electric field or pressure. And once again there is an instability that amplifies growth: when a new channel forms, it becomes a focus for the flow of surface water, producing further erosion.

But this does not mean that rivers can lengthen, and new branches form, only at the outermost tips, just as it is not so for electrical discharges or cracks. As in those cases, chance plays a role. All landscapes have random variations: in surface contours, soil type and permeability, rock type, vegetation cover, and so forth. This is the equivalent of the randomness of particle trajectories in diffusion-limited aggregation, say, and it ensures that there is always a chance that new tributaries may sprout downriver, on the higher-order streams of Horton's classification rather than only at the first-order stream heads. The chance is relatively smaller, because much of the water that falls on the ground already bounded by channels will be captured by them instead. But the possibility exists. So river networks grow like the other randomly branched patterns we have noted already: prey to randomness everywhere, but biased by preferential growth at the tips.

As Leopold and Langbein observed, drainage networks tend to be self-avoiding: stream heads hardly ever cut back across other streams to create islands or loops. This is because, as a stream head advances towards an existing channel, the area feeding it with water diminishes, since the other channel claims an increasing proportion of the water supply. Stream heads therefore generally run out of steam (or more properly, of water!) before they intersect other streams. Analogously, the tips of a DLA cluster very rarely intersect other branches because new particles cannot reach them once they get too close to another branch.

The connection between drainage network formation and crack formation is made explicit in a network-forming model called invasion percolation. Percolation is the passage of a fluid through a porous substance. David Wilkinson and J. F. Willemsen, working at the oil-mining company Schlumberger-Doll in Connecticut, devised the invasion percolation model in 1983 to describe how one fluid pushes another through a network of pores: a key process in oil recovery by injection of water into the oilfield. We have seen how this process can create branching instabilities that lead to viscous fingering patterns. But in invasion percolation the fluids are contained in a web of pores that imposes its own pattern, so that the invading fluid advances in a densely interweaving network (Fig. 4.4). In effect, the invading fluid pushes into a surrounding matrix in a way that is governed by the pores.

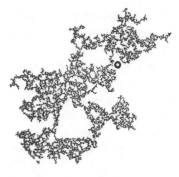

FIG. 4.4 Invasion percolation: the displacement of one fluid by another within a porous network. The 'invading' fluid is injected here at the point marked with a circle, and moves forward in a dense, convoluted system of loops and branches. (Image: Roland Lenormand, Institut Français du Petrole, Rueil-Malmaison.)

The probability that the invading fluid displaces the other depends on the size of the pore through which it passes, since this modifies the fluid pressure. If the pore network has a random distribution of pore sizes, then this probability varies more or less randomly through the system: there is an equal chance of a branch tip advancing at any point. The invasion percolation model captures this process by considering the invading fluid to be advancing as a 'cluster' through a lattice of obstacles linked together by bonds whose strength varies randomly from place to place. The fluid network grows by breaking these bonds and pushing between the obstacles, like flood waters breaking down barriers. The next bond to break is always assumed to be whichever is the weakest one around the perimeter of the cluster. You can now see that this model is very similar to the dielectric breakdown model described in the previous chapter, except that there is no ambiguity about which is the next bond to break: it is always the weakest, whereas in the dielectric breakdown model weaker bonds simply have a proportionately greater probability of breaking.

The advance of an invasion percolation cluster occurs mostly at the tips, because the rules of bond breakage mean that the cluster naturally 'seeks out' the weakest bonds in its path and leaves behind along its perimeter those bonds that are stronger. The chance of finding a bond at the tips that is weaker than those still unbroken deeper inside the cluster is usually pretty good; only rarely will all the tips happen to alight on strong bonds, forcing the breakage of one further back down the cluster's branches. The cluster grows into fractal form.

The British geomorphologist Colin Stark has proposed that the evolution of drainage networks is rather like invasion percolation. The breaking of bonds mimics the erosion of bedrock by a steady supply of surface water from rainfall; and the randomness in bond strengths reflects the non-uniformity of the landscape. To apply the model to this situation, he imposed one extra constraint—self-avoidance—so that a stream head may not intersect an existing channel. Stark showed that this model produced stream networks that looked rather realistic (Fig. 4.5). And he found that his model networks obey Hack's scaling law with an exponent of about 0.56, matching the value of 0.5–0.6 seen in nature.

However, like all simple models that have been proposed for explaining the form of river networks, the invasion percolation model only gets part of the pattern right. For one thing, snapping bonds in a lattice is not really much like erosion and sediment transport in real rivers. And sometimes three or more tributaries converge in Stark's networks, whereas this rarely happens in real river networks. Furthermore, it is a little unsatisfying that self-avoidance must be imposed rather than arising naturally in the model. In other words, there is more going on than mere invasion percolation in the formation of a river network. Nature tells us this must be so, because, like trees, river patterns do not all look the same.

Several alternative models offer a similar combination of landscape randomness and growth instabilities that promote branching, and almost all of

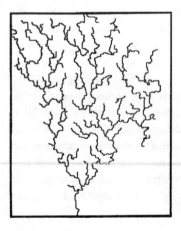

FIG. 4.5   The self-avoiding invasion percolation model of river network formation produces networks resembling those carved out as rivers cut back into bedrock. (After Stark, 1991.)

them produce fractal patterns, along with fair agreement with some of the scaling laws seen in the natural networks. On this evidence, then, it seems there is no unique way to describe the formation of river networks. We know the basic principles, but it is not clear which details are essential and which are incidental. And how well a model mimics the real world may depend on which rivers we choose for the comparison. This is all rather unsatisfactory.

THE BEST OF ALL WORLDS

What is needed is a broader perspective. Instead of a river winding through the landscape, think for a moment about an apparently simpler problem: the path of an object falling under the influence of gravity. It could be a cannonball fired from a cannon, or a pen rolling off a tabletop, or a raindrop released from a cloud. How can we calculate the trajectory? Newton's laws of motion provide the rules, telling us which forces act on the body. But these laws rarely supply a very easy prescription for calculating the trajectories that result. A better way of determining the paths of moving bodies was devised in the late eighteenth century by the French mathematician Joseph Louis Lagrange, and modified half a century later by the Irish mathematician William Hamilton. Lagrangian and Hamiltonian mechanics are equivalent to Newtonian mechanics, but instead of formulating the problem in terms of forces, they consider the energies involved—for example, the changes in the gravitational potential energy as an object falls, and its kinetic energy due to motion. This makes the job easier, and it provides a general criterion for the trajectory of a falling object, or indeed of any moving object: the path taken is that which minimizes a quantity called the action, which depends on the energy changes involved in the motion and the time taken for it to happen.

The Venezuelan environmental engineer Ignacio Rodriguez-Iturbe, the Italian physicist Andrea Rinaldo, and their colleagues think that there is an analogous 'minimization' principle guiding a natural river drainage network into a branched, fractal structure. The network evolves, they say, in such a way as to minimize the rate at which the mechanical potential energy of the water flowing through the network is expended. This claim needs a little unpacking.

As water flows downhill through a river network, it loses potential energy just as a falling cricket ball does. This is converted mostly into kinetic energy: the water moves. And the kinetic energy ultimately drives the process of erosion that leads the network to expand and rearrange its course. So the potential energy is ultimately *dissipated*: some of it goes into kinetic energy of the river water discharging into the sea (or at least, out of the drainage basin), while some is lost as frictional heat by the wearing away of rock and soil.

Suppose we had a divine ability to measure everywhere at once the amount of potential energy that all the water was losing each second. (We cannot hope to do this in real river systems, but it is something that can be easily totted up in computer models.) Rodriguez-Iturbe's minimization principle says that the network will evolve until it acquires the shape for which the total rate of potential energy dissipation is as small as possible.

It was not a totally new notion. The analysis of river drainage patterns conducted by Luna Leopold and his co-workers in the 1960s led him to conclude that these networks represent an optimal compromise between two opposing tendencies: for the expenditure of power by water flow to be as small as possible (which is more or less equivalent to Rodriguez-Iturbe's minimal energy dissipation), and for the power of the flow to be distributed more or less equally throughout the system. Leopold suggested that the river network evolves into a state where these things are balanced.*

To investigate what sort of network their model generates, Rodriguez-Iturbe and his colleagues 'evolved' a network created purely at random into one that obeyed their minimization principle; that is to say, they started with a network drawn on a checkerboard grid, in which water falling uniformly on each grid cell was drained along channels connected randomly from cell to cell (with the proviso that channels could not cross) into a network that ultimately flowed out of one corner. Then they rearranged the channels one cell at a time, calculating at each step how this affected the total rate of energy dissipation. Each change was made at random, but was 'accepted' only if it produced a decrease in this total

---

*Leopold insisted only that this optimal compromise be reached at the local scale, however—that is, for each segment of a network. He did not consider that the balance must also be achieved throughout the whole river basin. This is rather like saying that the network is like a collection of little 'houses of cards', rather than one delicately poised big one.

rate—if not, the change was discarded and another one was tried. This process can generate a wide range of networks, and each 'run' will create a different one. But all of them turn out to look realistic: they have scaling properties that obey Horton's laws, Hack's law, and several other empirical laws of river patterns too. Rodriguez-Iturbe and colleagues calls this set of possible solutions 'optimal channel networks'.

This seemed to suggest that the *form* of the network was indeed governed by the rule that energy dissipation from flow and erosion would be kept as small as possible: in this sense, the water flow will search out the 'best' solution. But can that really be so? It would be extraordinary if it was. If you think of a typical river running over a hilly landscape, the number of possible routes it might take is astronomical. The chance that the single 'best' route will be found is therefore utterly tiny, and indeed this is not really what happens. Instead, nature finds river network structures that are simply 'good enough': they might not be the best possible, but they are each the best in the neighbourhood. In an entirely analogous way, water flowing down a mountain range sometimes gathers in mountain lakes: the water could lower its energy still further if it found a way out of the lake and right down to the valley floor, but it cannot easily do that, and so gets stuck in the 'local minimum' of the lake. There are many alternative networks with broadly similar shapes that correspond to these local minima of the energy dissipation rate, and they are all, in a parochial rather than a global sense, optimal channel networks.

Why should river networks 'seek' to minimize their energy expenditure in the first place? Initially, Rodriguez-Iturbe and colleagues merely assumed that they did, and showed that this assumption gave realistic branching patterns. They did not initially attempt to justify the assumption. Kevin Sinclair and Robin Ball of Cambridge University tried to explain how the energy-minimization principle arises from the fundamental flow and erosion processes that govern network evolution. They deduced that the mathematical equations relating the water flow rate and the amount of erosion look like those that appear in Hamilton's law of least action. In other words, within the flow of a single stream of water over open ground lies the prescription for the shape of the Amazon. But of course one could never guess at that pattern by staking out a lone channel with any number of flow meters, depth gauges and so forth. The branching pattern is what emerges when that process is reiterated everywhere at once. And no two patterns are alike: they are all encoded in the physics of flow and erosion, but not uniquely defined by it.

Rinaldo, Rodriguez-Iturbe and their colleagues went on to make this connection more concrete by finding a completely different way to generate optimal channel networks, based on the erosional processes that happen in real river flow. They considered what would happen as rain falls uniformly onto a high plain whose roughness is totally random. A surface of this sort is more like sandpaper than a mountain range: it is uneven, but has no big peaks or deep valleys. The resulting flows of water create erosion that depends, at each point, both on the rate of flow and the steepness of the gradient down which the water runs. Again, the researchers divided up this conceptual landscape into a grid of cells, and assumed that the water would run out of each cell down the steepest gradient it can find. This flow can in principle produce erosion of the surface, but in the model it was assumed to remove and carry away material only when the flow exceeds some critical threshold value. When this happens, the height of the landscape at that point is reduced, altering the slope.

Notice that there is nothing in this model to ensure that the flow gets channelled into a single, connected river network. Yet as the simulation of landscape erosion proceeds, that is what happens (Fig. 4.6). And these networks turn out to have the same scaling properties and fractal dimension as the optimal channel networks produced by the earlier model, which in turn are excellent mimics of nature. So a set of purely 'local' rules that determine how the network and landscape evolve, describing the kinds of processes that happen in the real world, is sufficient to ensure that the river network finds its way to the 'optimal' shape that minimizes total energy expenditure.

What does the erosional process do to the shape of the landscape itself? I'll come back to this shortly.

IT'S SEDIMENTARY

Self-avoidance is the rule for river networks: the channels do not intersect. But when rivers flow across very flat, broad beds, they often break up into a series of channels that split and rejoin into a series of loops surrounding

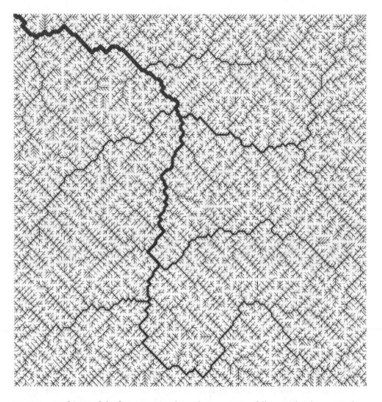

FIG. 4.6 In this model of river network evolution, water falls on a landscape with a randomly rough surface and causes erosion as it flows away. The resulting streams organize themselves into a joined-up river network in which the total rate of energy dissipation is a local minimum: it has the smallest value of all the networks with similar configurations. These so-called optimal channel networks have the same scaling laws as those seen in nature. (Image: Andrea Rinaldo, University of Padova.)

islands (Fig. 4.7). These are called braided rivers. Analogous skeins of water form on a smaller scale when streams run into the sea across a flat, sandy beach. The dried-up imprint of what appear to be braided rivers have been seen on Mars. The pattern appears whenever a broad sheet of water runs over a gently sloping, grainy sediment.

FIG. 4.7 Braided rivers have channels that loop and converge, creating isolated islands that form and vanish as the channels change their course. This one is the Waimakariri river in Canterbury, New Zealand. (Photo: Phillip Capper.)

'The sediments are a sort of epic poem of the Earth', says the ecological writer Rachel Carson—to which geologist Chris Paola of the University of Minnesota adds the frank confession that 'Unfortunately, this poem is written in a language we don't understand.' But he is one of those working on decoding it. Paola and his colleague Brad Murray have proposed, for example, that the transport of suspended sediment is crucial to the formation of braided rivers. Water scours out sediment in some parts of the flow and re-deposits it elsewhere to create sandbars and islands. If the rate of sediment removal by scouring increases sharply as the flow gets faster, then this sets up a positive feedback that makes a dip in the river bed get ever deeper: the dip captures more of the flow than the surrounding regions, and so proportionately more sediment is washed away from it. The reverse is true for a bump: the flow passes around it rather than over it, and so it suffers less erosion and gets higher than its surroundings. As a result, random small protrusions become islands that divert the flow to either side.

Murray and Paola devised a model of flow and erosion that captured these features. Water flowed across a checkerboard lattice of square cells, whose heights decreased steadily in one direction to produce a downhill flow. Superimposed on this smooth slope were small, random variations in height from cell to cell. The amount of water flowing through each cell depended on its height in relation to its uphill neighbours: the lower the cell, the greater its share of water from those uphill. The researchers assigned rules that governed how the amount of sediment either eroded or deposited at a cell depended on the flow across it. They found that their model results (Fig. 4.8) captured many of the features of real braided rivers. Channels continually form and reform, migrate, split and rejoin: the shape of the river is never steady. Although on average the flow of water and sediment down the river remains constant, it fluctuates strongly—more so than in non-braided rivers—because of this constant reorganization of the flow paths.

Any sandy sea shore will show you that shallow water flowing over grains is a rich source of pattern. When the water's edge retreats as waves lap at the beach, the sand can become moulded into diverse structures; the same patterns may be seen in the sediment at the bottom of an emptying reservoir. Adrian Daerr at the Université Pierre et Marie Curie in Paris and his co-workers studied these patterns in the lab in 2002, using a 'model shore' consisting of a plastic plate covered in a thin layer of sediment (alumina powder, mimicking sand or mud) and withdrawn slowly from a tank of water at various angles of inclination. The researchers found that as the angle was made progressively steeper, there were rather sudden changes in the kind of erosion pattern formed, from a densely 'cross-hatched' texture to branched channels to a dimpled 'orange-skin' surface and then to chevrons that overlap like fish scales (Fig. 4.9). They could not explain all these patterns, but suggested that the chevrons came about as a result of avalanches of grains down the slope of the sediment layer. Each avalanche spreads into a chevron, funnelling the flow of water in a way that sharpens the V shapes by erosion.

Such studies raise at least as many questions as they answer. But all these erosion patterns are self-organizing, in the sense that the flow becomes organized into a stable state with properties that remain statistically constant even though the details are for ever changing. This is a

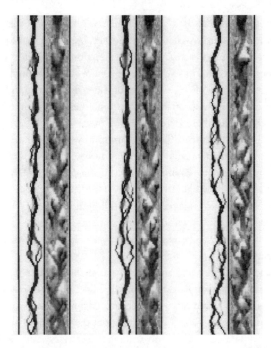

FIG. 4.8   A computer model of fluid flow and sediment transport captures the essential features of braided rivers. The pattern is constantly shifting, as shown here in three snapshots from a model run. The images on the right show the topography, and those on the left show the water discharge—essentially, the river channels themselves. (Image: Chris Paola, University of Minnesota.)

hallmark of self-similar growth, which allows an object to preserve its form while it grows indefinitely.

WHAT'S LEFT

When we think of river patterns, what usually comes to mind is the plan view: the convergent, branched network as seen from above, which Leonardo intuited and topographic maps and aerial photographs have now rendered familiar. But these images do not really reflect our

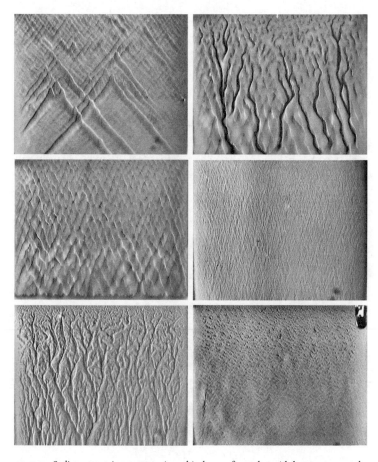

FIG. 4.9 Sediment erosion patterns in a thin layer of powder withdrawn at an angle from immersion in water. The pattern depends on the withdrawal rate and angle. (Photos: Adrian Daerr, ESPCI, Paris.)

experience of rivers, for what we see instead from our nose-high view of the world is the effect that the flow has on the landscape; in other words, we see the rugged profile that the river carves into the landscape: hills and valleys, gorges, ravines, and lone peaks (Plate 7). There is a characteristic shape and form in what the river *leaves behind*, just as there is in the course it takes.

While the river network is traced out as a pattern of wiggly lines, the topographic profile of the watershed is a *surface*. And just as the fractal nature of the channel network makes it more than a one-dimensional line, so too the rough erosion surface is a fractal that partially fills up three-dimensional space and so has a fractal dimension greater than 2. It is not hard to tell when a landscape is fractal, because you will soon find that it takes a lot more time and effort to travel between two points separated by a certain distance as the crow flies than it would on a flat plain. Journeys in fractal land are arduous.

What exactly does it mean, though, for a surface to be fractal? Simply put, it means that the bumps have no characteristic size scale: they come in all sizes. Put another way, it means that the apparent area of the surface depends on the size of the ruler that one uses to measure it. Consider getting from A to B over such a bumpy surface (Fig. 4.10). How far do you

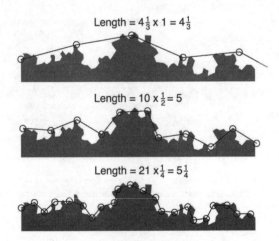

Length = $4\frac{1}{3}$ × 1 = $4\frac{1}{3}$

Length = 10 × $\frac{1}{2}$ = 5

Length = 21 × $\frac{1}{4}$ = $5\frac{1}{4}$

FIG. 4.10 A fractal boundary, such as a cross-section through a fractal surface, has a length that depends on the yardstick used to measure it. As the measuring stick becomes smaller, the apparent length seems to increase as we capture more of the details. Here the measured length increases slightly each time the measuring stick is halved in length.

travel? That depends on how you measure the route: smaller and smaller 'yardsticks' capture more and more of the ups and downs and so the overall measured length gets longer. Of course, the 'real' length doesn't get any longer—you just 'see' more of it. But the spooky fact is that, on a genuinely fractal landscape, there is no 'real' length at all. Ups and downs exist on all length scales down to the infinitely small, so the apparent length goes right on increasing as we measure with ever smaller rulers. As we have seen, true fractals like this are just mathematical abstractions, since the crenellations of any real boundary cannot get any smaller than the sizes of atoms. But some physical objects can remain fractal over a wide range of scales of magnification.

It was this apparent dependence of a rough object's perimeter length on the size of the yardstick that led Benoit Mandelbrot to uncover fractal geometry. In 1961 he came across the attempts of the English physicist Lewis Fry Richardson to specify the length of wiggly coastlines and borders, including the west coast of Britain and the border of Spain and Portugal. Richardson was interested in the causes of war and international conflict, and he was drawn to the issue of border length by thinking about how territorial disputes arise. He found that the apparent lengths of such boundaries measured from maps depended on the scale of the map that one used: small-scale maps show more detail than large-scale ones, and so capture more of the nooks and crannies, making the total length seem longer. Mandelbrot realized that 'length' is in this case not something that can be meaningfully specified. Instead, the 'shape' of the boundary is better described by calculating how quickly the apparent length increases as the yardstick gets shorter. This is what the fractal dimension measures: it is an unchanging geometrical property of the way the convoluted object fills up space.

There is an important subtlety here. I explained earlier that a fractal object such as a crack or a DLA cluster is self-similar: if you look more closely at any part of it, you seem to see the same overall shape repeated again. More precisely, self-similar objects are composed of (approximate) copies of themselves scaled down by a constant ratio. They have the same fractal dimension in all directions.

Fractal surfaces are not quite like this. Although they have a fractal dimension of between 2 and 3, indicating that they partially fill up three-dimensional space, this space-filling is not the same in all directions. We can see this rather easily by taking cross-sectional slices through a rugged mountainous landscape. A vertical cut reveals a rising and plunging, but continuous, transect across valleys and peaks (Fig. 4.11*a*). But a horizontal cut reveals something else entirely: isolated 'islands' separated by empty space (Fig. 4.11*b*). Such a fractal structure is said to be *self-affine*, which crudely means that the ratio by which the elements get scaled down at successive levels of magnification is different in different directions.

Mandelbrot realized in the 1970s that the natural topography of the Earth is typically a self-affine fractal. This was implicit, he said, in a description of mountain landscapes by the Victorian climber and explorer Edward Whymper in his book *Scrambles Amongst the Alps*: 'It is worthy of remark that... fragments of... rock... often present the characteristic forms of the cliffs from which they have been broken.' Self-affine landscapes can be generated rather easily on computers, and they supply convincing imitations of real mountainscapes (Fig. 4.12) which have been used in Hollywood movies such as *Star Trek II: The Wrath of Khan*. The crucial point here is that these landscapes are not simply random. If you set the computer to generate an image in which the size and shape of the hills and valleys are assigned by a random-number generator, the result is a topography that is certainly uneven but looks wrong. Fractal

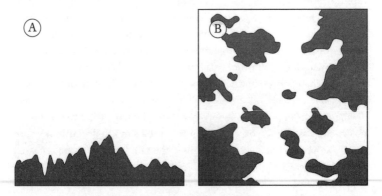

FIG. 4.11   A vertical slice through a rugged valley reveals an irregular profile of peaks and valleys (*a*). A horizontal cut, meanwhile, isolates islands—contour lines corresponding to cross-sections of the peaks, separated by gaps (*b*).

FIG. 4.12 Self-affine fractal landscapes generated by computer look very like real mountainous terrain. (Image: John Beale.)

landscapes are 'noisy' and unpredictable, but there is more to them than chance alone.

In view of how important fractal coastlines were to the evolution of the general notion of fractal geometry, it's surprising that a good model for how coasts acquire this form by erosion was not developed until 2004. Partly this is due to the fact that coastal erosion is a complex, many-facetted process. The pounding of waves can wear away rock, but it also redistributes sand and stones. Cliff faces are shattered by frost expansion or worn down by rain. And water induces chemical attrition called weathering, for example by salt corrosion or via reactions between minerals and water. Rocks weakened by slow chemical weathering may be broken apart and removed quite suddenly by the violence of a storm.

Bernard Sapoval at the Ecole Polytechnique in Palaiseau, France, and his co-workers simplified these details by assuming just two kinds of erosion: rapid, due to the mechanical effects of battering in storms, and slow, due to chemical weathering. In their model, a coast was represented by a grid of cells containing several rock types with different resistance to erosion, distributed at random. They suggested that the key to the evolution of fractal geometry was a feedback process by which erosion of the coast by waves is altered as the coastline becomes more convoluted, damping down the waves. Allowing for this effect, the researchers found that a smooth coast soon breaks up into a rough outline, which becomes increasingly pitted into bays, peninsulas, and islands (Fig. 4.13). The erosion process doesn't just expose the strongest rock—this would result in a random, non-fractal shape—but maintains a fine balance between removing the weaker sections and removing those that are most exposed. The model ignores

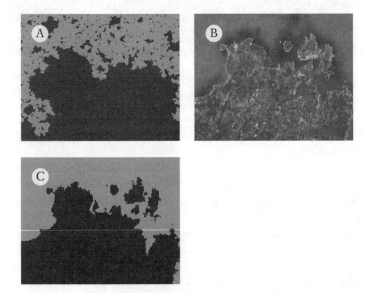

FIG. 4.13   A coastline carved out by a computer model of erosion (a: the dark area here is the land) looks similar to a real coast (b: in Sardinia; the boundary is shown more clearly in c), with peninsulas, bays, and a ragged outline. Both boundaries are fractal, with the same fractal dimension. (Images: Bernard Sapoval, CNRS Ecole Polytechnique, Palaiseau.)

several aspects of coastline formation that are surely important, not least the movement of sediment, but nonetheless it seems to capture the most obvious features of real coasts. They 'look' right, for sure.

CARRIED AWAY

The self-affine relief of a river basin must be related to the branched structure of the river network. But how? Ignacio Rodriguez-Iturbe, Andrea Rinaldo, and their co-workers think that their model of optimal channel networks provides a link. I showed earlier that in this model, water carves realistic-looking river networks through a randomly bumpy surface. In doing so, the flow reshapes the topology of the landscape, sculpting it into peaks and valleys. The initially sandpaper-like surface deepens into a rugged range of hills and valleys with a fractal character (Fig. 4.14), very like that of a real landscape. Once this landscape has attained its fractal state, erosion continues and the shape continues to change; but the basic form, as characterized by the fractal dimension, remains constant.

Tamás Vicsek and his co-workers in Budapest have studied this same process experimentally, using real mud and water. They mixed sand and soil to simulate the granular, sticky stuff of hillslopes, from which they constructed a flat-topped ridge just over half a metre long. They sprayed it evenly with water to see what kind of surface would be carved out by erosion.

The running water carries off material in two ways: the granular substance is worn down gradually, but from time to time landslides remodel the surface more abruptly. Both of these processes, of course, occur in real hill and mountain ranges. The result is a rough, bumpy ridge that one could easily mistake for a rocky hillslope on a scale thousands of times bigger (Fig. 4.15)—a reflection of the scale-invariant self-affinity of these erosion surfaces. But you might protest that mountains are made of rock, not a soft mixture of soil and sand. That is true, but it may not matter. Both of these substances are worn away by flowing water; it merely happens much faster in the softer medium. And both have a resistance to erosion that varies rather randomly from place to place: the sand and soil were only crudely mixed, and rock is highly non-uniform. Finally, both

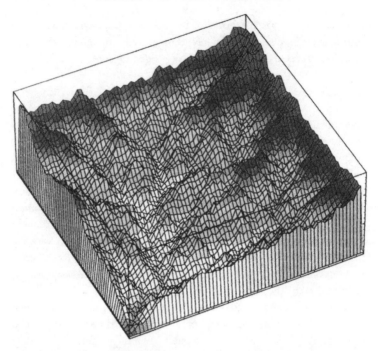

FIG. 4.14 In the optimal channel network model described earlier, a river network such as that in Fig. 4.6 has an associated topography. The initial landscape is randomly bumpy, but its profile changes to a fractal form, with hills and valleys of all sizes and heights. (Image: Andrea Rinaldo, University of Padova.)

materials suffer erosion due to the same two processes: gradual removal of suspended small particles, and abrupt landslides.

The results, then, are miniature replicas of the world's most awesome vistas: demonstrations of how the elemental forces of nature can exhibit a blithe indifference to scale.

INVERTED ICICLES

The snowfields of the Andes experience a very different kind of erosion process that creates one of nature's strangest spectacles. The high glaciers

FIG. 4.15 An experiment on erosion of a bed of sand and clay by water produces a rugged skyline (a) that resembles those seen in nature at scales thousands of times larger, such as this mountainscape in the Dolomites (b). (Photos: Tamás Vicsek, Eötvös Loránd University, Budapest.)

here can become moulded into a forest of ice spires, typically between 1 and 4 metres high, called penitentes because of their resemblance to a throng of white-hooded monks (Fig. 4.16a). Charles Darwin saw these eerie formations in 1835 en route from Chile to Argentina. 'In the valleys there were several broad fields of perpetual snow', he wrote in *The Voyage of the Beagle*,

> These frozen masses, during the process of thawing, had in some parts been converted into pinnacles or columns, which, as they were high and close together, made it difficult for the cargo mules to pass. On one of these columns of ice, a frozen horse was sticking as on a pedestal, but with its hind legs straight up in the air. The animal,

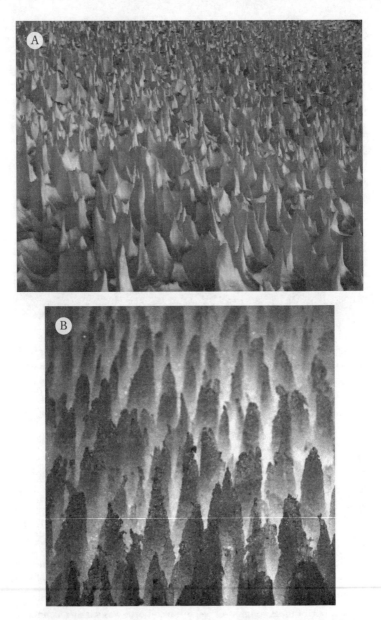

FIG. 4.16 Ice and snow fields in the Andes can be carved into spikes called penitentes by the eroding power of sunlight (a). The process has been mimicked in the laboratory, producing ice spikes just a few centimetres tall (b). (Photos: a, Cristian Ordenes; b, Vance Bergeron, Ecole Normale Supérieure, Lyons).

I suppose, must have fallen with its head downward into a hole, when the snow was continuous, and afterwards the surrounding parts must have been removed by the thaw.

Darwin remarked that the locals believed them to be formed by wind erosion. But the process is more complicated than that, representing a classic case of pattern formation by self-amplifying feedback. The air at these great heights is so dry that sunlight falling on the ice transforms it straight into water vapour rather than melting it into liquid water. A small dimple that forms in the smooth ice surface by evaporation acts as a kind of lens that focuses the sun's rays into the centre, and so it is excavated more quickly than the surrounding ice. It's a little like diffusion-limited aggregation or dendritic growth in reverse: a 'fingering' instability penetrates into the ice rather than pushing outwards from the surface.

The process can be accelerated by a fine coating of dirt on the snow surface. As the troughs deepen they expose clean snow that is prone to further evaporation, whereas dirt in the old snow at the peaks covers the ice crystals like a cap and insulates them. You might expect that, on the contrary, snow or ice will melt faster when dirty than when clean, because the darker material will absorb more sunlight. But whether a layer of dirt acts primarily as an insulator or an absorber depends on how thick it is.

The physicist Vance Bergeron of the Ecole Normale Supérieure in Lyons and his co-workers have mimicked this natural process in the laboratory, making 'mini-penitentes' by exposing blocks of snow or ice to a bright spotlight. After a few hours of illumination, tiny peaks just a few centimetres tall appeared in the ice (Fig. 4.16b). Structures this small are found naturally, and are thought to be the precursors of full-scale penitentes. It is possible that this process might be exploited on still smaller scales, using fierce lasers in place of sunlight to erode the surfaces of silicon wafers used in solar cells. A coating of microscopic penitente-like peaks could make the silicon less reflective, so that it can capture more of the sunlight that falls on it. That is the wonder of these pattern-forming processes: they may operate over such an immense range of scales, from mountains to molehills to the microworld, and with such indifference to context, that we can find geology inspiring technology, or biology apeing a snowflake.

TREE AND LEAF

Branches in Biology

The tree (*dendros*) metaphor is invoked so regularly in scientific descriptions of branching patterns that it seems only reasonable to expect these models and theories to tell us something about the shapes of real trees. But therein lies a problem of another order. A tree is a teleological form: it is a form with a 'purpose', an example of Darwinian designer-less design.

There are many challenges that a tree must meet. How can it pump water from the roots to the leaves? How can it support its own tremendous weight? How to maximize its light-gathering efficiency? How to grow tall enough to compete for light with its neighbours, without becoming too massive for the roots to bear? In the face of these dilemmas, there is little chance that a simple model based in maths or physics will tell all about the shape of a tree. It is far from clear, for example, that this shape is dictated by anything as simple as a branching growth instability.

If you were to be asked to describe the shape of a planet or a grain of salt, you can do so in a single word: 'sphere', or 'cube'. But such geometric labels do not work for trees. 'Branched' is clearly not enough; indeed there is no generic 'tree shape' anyway, since this varies between species (Fig. 5.1 and Plate 8). To give a precise description you would need to specify all of the branches and all of their angles and lengths—to paint in words a

FIG. 5.1 The branching patterns of trees are a fingerprint of their species. (Photos: a, Henry Brett; b, Kyle Flood; c, Amanda Slater; d, Andrew Storms).

picture of the complete tree (and then only of that particular tree). You end up, in other words, like Sartre's Antoine Roquentin in *La Nausée*, horribly fixated on the arboreal specifics in front of you.

The most useful kind of mathematical description of a tree doesn't do this. Instead, it encapsulates a general shape in a set of rules—an *algorithm*—that generates a whole family of characteristic yet non-unique forms.

One might devise a 'cypress algorithm', say, or an 'oak algorithm'. An algorithmic approach to generic form underlies much of the work on mathematical fractals: the algorithm states how to 'grow' the form by enacting a series of steps. It is important to remember that there's no guarantee that the mathematical algorithm bears any relation to the growth process that occurs in nature; but on the other hand, it does seem clear that some natural forms are created by the repeated implementation of certain simple steps.

Algorithmic models are capable of mimicking the essential shapes of trees without paying the slightest regard to the biology or mechanics that underpins them. These models are not descriptions of the real growth process, but rather prescriptions for enabling us to generate something tree-like (or plant-like). Even if that does not tell us much about the reasons why a tree looks the way it does, it might provide clues to the primary characteristics of this form, so that we can begin to make educated guesses about what a true model of growth should look like.

Leonardo da Vinci suspected (although without formulating it in quite these terms) that there are algorithmic rules governing tree growth. For example, he suggested that at branching points, the rule is that the central trunk is bent by some specific angle when a side branch occurs on its own, but is not bent at all if two side branches are positioned opposite one another. An artist could regard these simply as rules of thumb for making his paintings look realistic; but as I explained in Book II, for Leonardo representing nature was intimately bound up with understanding it. He felt that the crucial aspect of a branching junction is that the total cross-sectional area of the branches stays the same. As a hydraulic engineer, he considered it important that these 'pipes'—for the wood of a tree is a mesh of cellular channels, the phloem and xylem, that carry water and sugar-rich liquid to and from the extremities—retain their fluid-bearing capacity as they branch.

Are Leonardo's rules for branching angles true? Yes, to a degree (Fig. 5.2), but they seem to depend on the size of the side branch: single small ones cause next to no bending of the trunk. The German anatomist and embryologist Wilhelm Roux, a pupil of Ernst Haeckel at Jena, attempted to specify these rules more precisely at the end of the nineteenth century. He claimed that:

FIG. 5.2 Tree branches often tend to follow Leonardo's rules that a single side-branch deflects the main trunk (a) while two, on opposite sides of the trunk, do not (b).

1. When the central stem forks into two branches with equal width, they both make the same angle with the original stem.

2. If one branch of the fork is of lesser width than the other, then the thinner branch diverges at a larger angle than the thicker.

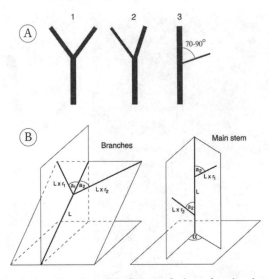

FIG. 5.3 Rules for branching. One of the first sets of rules to formalize the way natural branching occurs was drawn up in the nineteenth century by Wilhelm Roux, who formulated them for the cardiovascular network (*a*). Hisao Honda devised some rules for creating realistic-looking tree patterns in an algorithmic way (*b*). The rules illustrated on the left apply to all branches except those that diverge from the main trunk, which are governed by the rules on the right. Here r_1 and r_2 are the ratios of the lengths of side-branches to that of the main branch or trunk (*L*), and the angles a_1 and a_2 specify the respective divergence angles. Successive branches off the main trunk diverge from one another by an angle α.

3. Side branches small enough that they do not deflect the main stem appreciably diverge at angles between 70° and 90°.

These rules are illustrated in Fig. 5.3*a*. Roux did not assert them for trees, however: he was studying arterial networks, and it was only in the 1920s that the biologist Cecil Murray suggested they applied to plant stems. Murray too was more interested in blood flow, but researchers like him had started to suspect that the similarity in form between trees and the blood circulatory system might be no coincidence. Both are *vascular* networks: they are formed from hollow tubes that carry fluids. Veins and arteries are like pipes of varying width, while wood is like a bundle of narrow tubes that separates into bunches at branching points.

Murray's algorithmic rules generate somewhat realistic-looking 'trees' when they are used algorithmically to create a randomly branched network. A more complex algorithm for making tree-like branching structures was proposed by the Japanese biologist Hisao Honda in 1971, and runs as follows (see Fig. 5.3b):

1. Every branch forks into two 'daughter' branches at a single branching point.

2. The two daughter branches are shorter than the 'mother' branch by constant ratios r_1 and r_2.

3. The two daughter branches lie in the same plane as the mother branch (the branch plane), and diverge from it at constant angles a_1 and a_2.

4. The branch plane is always such that a line lying in this plane, perpendicular to the mother branch, is horizontal. (This is the trickiest of the rules to envisage, but is explained in the figure.)

5. An exception to (4) is made for branches diverging from the main trunk, which observe the length ratios specified in (2) but branch off individually at a constant angle a_2, with a fixed divergence angle (here denoted α) between consecutive branches.

With a few minor changes, this set of rules can be used to define algorithms that produce a whole range of branching patterns closely mimicking those of real trees (Fig. 5.4). Further modifications to account for the influences that real trees experience (wind, gravity, the need to arrange leaves for optimal light harvesting) give even more realism. Honda's algorithm is *deterministic*: it prescribes the branching pattern fully once the ratios and angles are fixed. Other algorithms that have been used to generate life-like trees in computer art employ random elements to create more irregular forms. In nature, randomness enters into the branching patterns as a consequence of such things as breakages, collisions between branches, growth stunting due to the shade of an overlying canopy, and the mechanical influences of wind and rain.

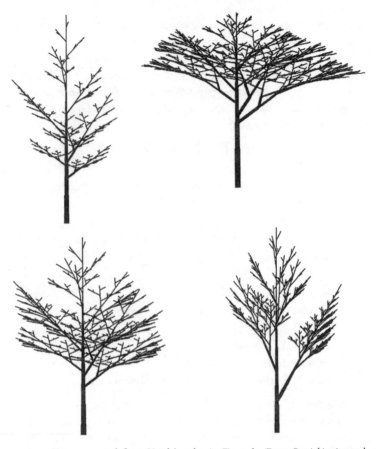

FIG. 5.4 Trees generated from Honda's rules in Fig. 5.3*b*. (From Prusinkiewicz and Lindenmayer, 1990.)

Another class of deterministic algorithms, called L-systems by Przemyslaw Prusinkiewicz of the University of Regina in Canada, can be used to produce plant- and fern-like structures (Fig. 5.5). Ultimately, one might hope that appropriate rules for these tree-growing algorithms can be *derived* from models of how plants grow, such as those mentioned in Book I.

PLATE 1: A snowflake displays a delicate balance of chance (in the initiation of branches) and determinism (the sixfold symmetry).

(Photo: Ken Libbrecht, California Institute of Technology)

PLATE 2: The wispy boundaries of many clouds trace out a fractal form.

(Photo: Maciej Szczepaniak)

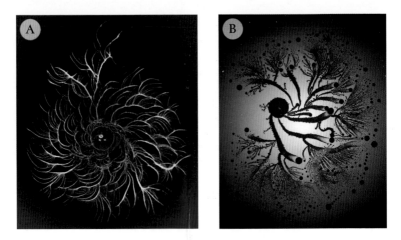

PLATE 3: Branching patterns in bacterial colonies.

(Images: Eshel Ben-Jacob and Kinneret Ben Knaan, Tel Aviv University)

PLATE 4: The entrance to Fingal's Cave on the island of Staffa in Scotland is announced by a colonnade of natural pillars of roughly hexagonal cross section.

(Photo: Lucas Goehring, University of Toronto)

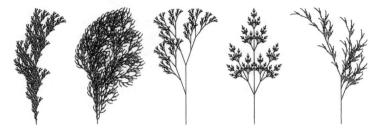

FIG. 5.5 Plants and ferns generated by 'deterministic' branching algorithms, where the pattern is completely specified by the rules (that is, it contains no random elements). The same motifs recur again and again at different scales in these structures, but regularity is evident to greater or lesser degrees. (From Prusinkiewicz and Lindenmayer, 1990.)

SCALING UP

When Cecil Murray posited his rules of vascular network shape in the 1920s, he went a crucial step further. Rather than merely describe these networks, he also wanted to explain why they have the form they do. He proposed that they embody a parsimonious *minimization* principle, entirely analogous to that which we encountered in the previous chapter for river networks. The vascular network, said Murray, has the form that requires the smallest amount of work to pump the fluid around it. He argued that the energy required to drive blood down branching arteries is smallest if narrow branches diverge at large angles and wide ones diverge at shallow angles. It seems clear that Leonardo da Vinci had something of this nature in mind when he formulated his own tree-branching rules—he too was thinking about what kind of branching geometries were best suited to efficient fluid flow through the system.

The analogies in form and function between rivers and biological branching networks as fluid-distributing systems led the geologist Luna Leopold in the 1970s to consider whether his conclusions about the former might be generalized to the latter. We observed in the previous chapter that Leopold suspected river networks were shaped by a balance between distributing the power of water flow evenly and ensuring that as little of this power is expended as possible. Might plants and trees be governed by similar criteria, he wondered—and if so, what are they? The equivalent of an even distribution of power, he suggested, was a uniform spreading-out of a plant's canopy so that it might best harness the available sunlight. And instead of minimizing the power expenditure of the flow, he thought that plants might seek to minimize the total length of the branches. It was a reasonable thing to suppose, given that building new wood or stem costs energy, although Leopold didn't bother too much about the fact that branches of the same length but different thickness contain very different amounts of tissue.

Wolfgang Schreiner of the Institute for Medical Computer Research in Vienna, Austria, and his colleagues have shown that a wide variety of different 'optimal' branching patterns can be generated by growing them according to rules that minimize different quantities. They have considered networks to which new branches are added in succession, under the condition that each new branch, starting at some specified point, makes the connection that minimizes some quantity related to the width and length of all the existing channels. This sounds a little abstract, perhaps, and indeed it is: the researchers were not trying to specify *what* gets minimized, but simply to see what the patterns look like for different choices of this criterion, whether it be total network length, energy dissipation, or whatever. They found that different choices led to markedly different structures, all of them looking more or less familiar (Fig. 5.6). Some looked like meandering rivers on open plains, others like the streamlined drainage networks seen in mountainous terrain, others like the veinous patterns on leaves or the passageways of lungs. This does not tell us what controls the structure of any of these particular networks, but it suggests that this combination of algorithmic growth (repeatedly applying a set of rules) and minimization (making the 'best' connections) could contain the seeds of an explanation for a wide range of network structures.

But perhaps the most striking and provocative application of ideas about minimization to branching network shapes has come from a collaboration between two ecologists, James Brown and Brian Enquist from the University of New Mexico, and a physicist, Geoffrey West of the Los Alamos National Laboratory. They have proposed that the branched and fractal nature of fluid distribution systems in organisms can explain a longstanding biological puzzle about how life's processes depend on body mass.

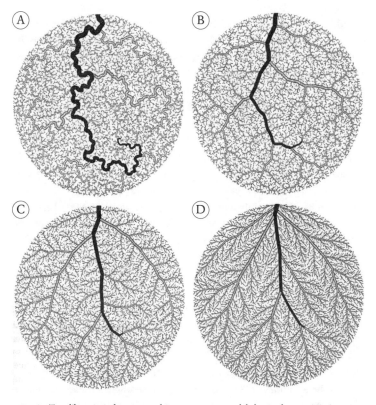

FIG. 5.6 Tree-like networks generated in a computer model that seeks to minimize some 'global' property of the system, specifically the total value of some aspect of all the branch segments. The network in *a* minimizes the total segment length, that in *b* the total segment area, in *c* the total volume, and in *d* a 'four-dimensional' property, the so-called hypervolume. In each case, the result is a branching pattern that looks familiar: *a* looks like a river on rather flat terrain, for example, and *c* and *d* like rivers in increasingly mountainous terrain. (From Schreiner *et al.*, in Brown and West (eds), 2000.)

Small creatures have faster heartbeat tempos than big ones: babies' hearts beat faster than those of adults (they also breath faster), and the heartbeats of small creatures such as birds are more rapid still. This relationship of beat rate to mass can be expressed in precise mathematical terms, and it turns out to be a power law, or scaling law (see page 38). For a wide variety of organisms, the heartbeat rate turns out to be proportional to the inverse of the body mass raised to the power $1/4$. The life span of an organism, meanwhile, is directly proportional to the body mass raised to the power $1/4$—the bigger they are, the longer they live. This means, incidentally, that the total number of heartbeats in an organism's lifetime (the heart beat rate multiplied by the life span) is a constant for all organisms, equal to about one and a half billion. Mouse, human or elephant: all have the same allotted time, when measured in heartbeats.

Anyone who knows how often children need feeding will also be familiar with the idea that smaller organisms have a faster metabolism. An organism's metabolic rate—the rate at which it consumes energy—is proportional to the $3/4$ power of body mass, which means that small creatures need more energy pound for pound: the shrew must consume more than its entire body mass every day. There are several other examples of these biological scaling laws: the cross-sectional area of aortal arteries in mammals and of tree trunks also depend on body mass raised to the power $3/4$, for instance. They are all called *allometric* scaling laws, and they are obeyed by organisms ranging from microbes to whales.

Of course, one would expect large creatures to use up more energy than small ones. But it is not at all obvious that the same scaling law should be followed over such a huge range of body sizes. Still more puzzling are the actual values of the powers in the scaling laws: they all seem to be multiples of $1/4$. If the biological parameters (heartbeat and so forth) were related to how quickly fluids could be distributed in the body, you would expect the relationship to depend on how far they have to travel, which would seem to be proportional to the body's dimensions. These in turn are proportional to the $1/3$ power of body mass.* You would therefore think that all these scaling laws should come with powers that are multiples of $1/3$, not $1/4$. It seems almost to imply that bodies are four-dimensional, not three-dimensional.

Enquist and colleagues suspected that the branched structure of the body's distribution networks might hold the key to understanding these

*This might be easier to see the other way around: the body mass is directly proportional to the body volume, which varies as the cube—the 3rd power—of the body's linear dimensions. So the time taken to travel at constant speed across a cube-shaped box depends on the $1/3$ power of the box's volume.

allometric scaling laws. They reasoned that ultimately organisms can survive only so long as their tissues are supplied with the essential resources that sustain them. Air-breathing creatures like us need a steady supply of oxygen to our cells. This is delivered to the blood via the branching channels of our lungs, and distributed around our bodies by our cardiovascular network. Trees and plants likewise need a vascular system to provide water and nutrients. But branching itself is not enough: the key point is that the branching is hierarchical, with big branches splitting into smaller ones and those into channels smaller still. This tends to produce fractal structures, which have the advantage that they exist 'between dimensions': they extend throughout all of the space they occupy without filling it entirely (which would leave no room for tissues themselves).

The researchers modelled these networks as systems of tubes which become progressively thinner at each branching point. There are two governing principles for the networks. First, all of them, regardless of size, have to end in tubes of the same size. These terminal branches can be considered to be the analogues of the smallest capillaries in cardiovascular systems, whose size is geared to that of the organism's individual cells—which varies little regardless of the total body size. Second—and here is the crucial point—the network is structured so that the amount of energy required to transport fluids through it is minimized.

For plant vascular systems, the passages of the network are in fact bundles of vessels of identical cross-section. At each branching point, the bundles split into thinner bundles with fewer vessels in each. For this situation, the researchers showed that the $3/4$ scaling law of metabolic rate with body mass falls out quite naturally from an analysis of the geometric properties of the energy-minimizing network. For mammalian distribution networks, on the other hand, the situation is rather more complex, and a $3/4$ scaling law is obtained only when the model includes the facts that the fluid flow is pulsed (due to the pumping of the heart) and the tubes are elastic. Most importantly, these relationships apply only for fractal distribution networks. Many human-made systems that distribute fluids or energy for power generation, such as combustion engines or electric motors, do not conduct this distribution through fractal networks, and they show power-law scaling with exponents related to $1/3$, not $1/4$, as their mass increases.

This cannot be the whole answer to allometric scaling laws—for one thing, they are obeyed by organisms that do not have branched distribution systems—but it implies an intriguing significance for fractal networks in the living world. James Brown suggests that it is in fact the ability of fractal networks to provide an optimal supply system to bodies of different sizes that enables living organisms to show such a huge range in body shapes and sizes, extending over 21 levels of magnification by ten from bacteria to whales. The theory also demands some rethinking of how biology works. The emergence of the science of genetics has encouraged the view that living systems from cells to ecosystems are governed primarily by the transfer of information: genes control body shapes and determine fitness, and populations grow and diversify owing to the shifting make-up of their genetic resources. But the work of Enquist and colleagues places the spotlight instead on *energy*: they say it is the need to distribute energy efficiently, with minimal waste (dissipation), which determines the patterns of the distribution network, with knock-on consequences for the size of creatures and thus for their abundance, their activity and their longevity. The researchers have argued that their theory can even explain biological properties at the level of ecosystems, such as how the density of plants or animals that occupy a patch of ground depends on the masses of the individual organisms. Such relationships also tend to follow scaling laws with exponents that can be explained by considering how much 'metabolism' can be sustained by the available energy. Enquist, Brown, and West believe that their scaling laws, derived from the branched networks of distribution systems, may provide a unified view of how the living world works.

WEBS OF LIFE

With his preference for answers to pattern and form that invoke engineering rather than 'black-box' Darwinism, D'Arcy Thompson would no doubt have understood and appreciated all of this. Indeed, he discussed biological scaling laws in the early pages of *On Growth and Form*, explaining for example how the mechanics of a tree dictate that, if it is not to bend under its own weight, the diameter of the trunk must increase in proportion to the height raised to the power $3/2$. 'Among animals we see how small birds and beasts are quick and agile, how slower and sedater movements come with larger size', he wrote.

Yet there is, as we have seen, another strand to Thompson's arguments that demands to know not just why some things are possible and some not, but how the pieces of the puzzle are put in place. In the same vein, so to speak, it is one thing to say that a fractal branching network offers optimal distribution of fluids and minimal energy dissipation; but the growing organism does not know that. It is clearly not the case that every little passage of the lungs is specified by a genetic blueprint, since no two lungs are alike, even in the same body, any more than you can find a sycamore tree that can be exactly superimposed on another. These networks have to be grown, and clearly that must happen via the now familiar combination of chance and necessity: the generic form of the structure, for example in terms of the fractal dimension and the average angles of junctions, does not vary significantly from one case to the next, but the details are always different. What are the rules that make this happen?

Consider, for example, the blood vessels of the cardiovascular system. One of the best studied subsystems is the network of blood vessels in the human retina (Fig. 5.7), which is a particularly dense-branching network, since the retina has the highest oxygen requirement of any tissue in the human body. The ophthalmologist Barry Masters at the University of Bern in Switzerland has collaborated with physicist Fereydoon Family in Atlanta, Georgia, to calculate the fractal dimension of this network, and they find that it has a value of around 1.7: the same as is seen in clusters of particles grown by diffusion-limited aggregation, and in cracks and electrical discharges formed in the dielectric breakdown model. These, we saw earlier, are examples of processes governed by so-called Laplacian growth, where positive feedbacks amplify small, random bumps at the growth front and turn them into new branches.

Does this mean that the growth of the retinal vascular network is similarly governed by Laplacian growth instabilities? That is possible, but it doesn't necessarily follow. For one thing, we have seen that the fractal dimension is rather a broad-brush characteristic of a branching structure: there is no reason to believe that only a single mechanism can give rise to a network with a fractal dimension of 1.7. And in any event, the

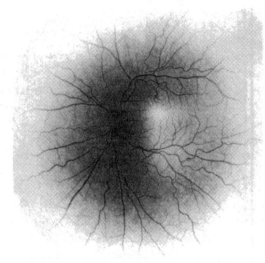

FIG. 5.7 The blood vessels around the retina form a fractal branching network with a fractal dimension of about 1.7. (Photo: Barry Masters, University of Bern, kindly provided by Fereydoon Family, Emory University.)

retinal vasculature doesn't look much like a DLA cluster: it is far less bumpy and convoluted. Moreover, the process by which growing blood vessels proliferate and develop branches, called angiogenesis, is complicated and doesn't always generate a diverging, randomly branched structure—often the vessels are interconnected in more complex ways. For example, blood vessels may sometimes intersect and join to form closed loops: they do not necessarily exhibit the self-avoidance we saw in river networks (and which is generally a property of DLA too). This is particularly evident in the vein networks of leaves and plants (Fig. 5.8), which grow via processes similar to angiogenesis. The reconnection between two branches in a vascular system is called anastomosis, and it means that there is more than one possible route for getting from one point to another in the network. So these vascular systems are more like the Paris metro than like a tree: if you want to go from A to B, you often have a choice of several possible paths.

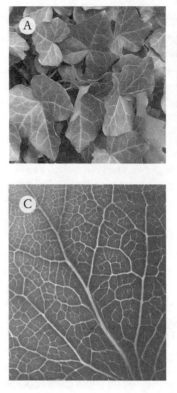

FIG. 5.8 The branching patterns of plant vascular systems: ivy leaves (*a*), a sea fan (*b*) and a part of the vein network in a leaf of the African shrub griffonia (*c*). (Photos: a, Lars Hammar; b, Gary Rinald; c, Peter Shanks.)

The first stage in the formation of a vascular network is the appearance of a web of vessels made from cells called angioblasts. This appears to happen by chemotaxis: the cells emit some chemical substance that diffuses into the surroundings and attracts others to move towards them, homing in on the densest concentration of the attractant. As we've seen chemotaxis enables bacteria to talk to one another and coordinate their movements, forming complex patterns and flows. Angioblasts emit a protein chemo-attractant called vascular endothelial growth factor, the influence of which causes them to gather into chainlike bunches that intersect in a web and eventually become veins.

143

The fractal, branched shape of this web emerges in the next step, which is angiogenesis. The tissues that the vascular network supplies with vital fluids get steadily bigger as the organism grows: the channels are transport networks in an expanding landscape. If any part of the landscape becomes too distant from one of these supply routes, it risks becoming depleted in life-sustaining ingredients (oxygen or nutrients, say), and a new route is needed. But growing a new vein, like building a new road, has a cost in energy and materials, and it cannot be done lightly. How are cells at risk identified, and how do they stimulate the formation of a vein? The form of the vascular network must ultimately derive from the answers to these questions.

Angiogenesis also depends on diffusing chemical signals. Tissue cells far from existing blood vessels begin to produce and emit proteins called angiogenic factors: such cells are said to be ischemic. When the chemical messengers reach a nearby vessel, it sprouts a new limb which grows in the direction in which the concentration of the chemical signal increases, that is, towards its source. In effect the distressed cells are saying 'throw me a line'. An analogous process produces the characteristic vein patterns of leaves (Fig. 5.8), where it seems that the chemical signal is supplied by the plant hormone auxin. I showed in Book I how auxin controls the appearance of new leaves or florets at the tip of a growing plant stem, giving rise to the remarkable mathematical regularities in the arrangements of these features on the plant in the process called phyllotaxis. In a nice echo of the fractal nature of many plants, it seems that the same substance dominates the formation of the branching pattern at the smaller scale of individual leaves.

The distribution of auxin throughout a leaf is often spotty: there are isolated sites that are rich in auxin, surrounded by regions where the concentration is lower. New veins will tend to grow towards these auxin spots, which the plant interprets as a plea for more nutrients. It is tempting to imagine that the auxin spots correspond to locations where auxin is being produced at a higher rate—in which case the shape of the network will be dictated by the underlying pattern of auxin spots. But it seems that this spotty pattern might instead be self-organized out of a tendency for the leaf cells to produce auxin at a constant rate everywhere.

Pavel Dimitrov and Steven Zucker at Yale University have shown how this may come about. They have represented a portion of leaf as a collection of cells that all produce and release a signalling hormone, denoted S. Each cell can 'detect' the concentration of S inside itself and in its neighbours. If an adjacent cell holds much less S, the original cell may alter its surface membrane so as to let out more S to that neighbour. Cells that increase their permeability this way are on the road to becoming vein cells, which conduct S more freely than ordinary cells. The formation of a nascent vein thus channels the hormone away from hotspots towards regions of low concentration. Since existing veins carry the hormone away efficiently, those low-concentration regions are typically such veins themselves—and so a channel grows between a hotspot and the nearest vein. The more distant a patch of leaf is from its nearest veins, the more S is likely to accumulate and the stronger is the signal promoting the formation of a new vein leading to that site. This process is a little like the way river networks form: rain falls equally everywhere, but the water runs down the steepest gradient until it reaches a river, gradually carving out a new tributary.

This model generates the kind of vein growth patterns seen in real leaves; in particular, new veins tend to join existing ones at right angles. Veins may appear as 'stand-alone' branches, or they may form junctions and loops. As the network evolves on a growing grid of cells, the hotspots of S move around: old ones are 'relieved', while new areas become starved of access channels. The auxin spots of a growing leaf are similarly ephemeral, suggesting that there is not after all some underlying blueprint for where the branches should go. The pattern draws itself, each stage setting the scene for the next.

This is comparable to the way a combination of diffusion and reaction of chemical substances within a uniform matrix can give rise to spontaneous pattern formation: the process I described in Book I, where I showed that it can account for patterns like the spots, stripes, and networks seen on animal skins. Hans Meinhardt, a German biologist who has pioneered the understanding of these so-called chemical reaction–diffusion systems and of how they might produce biological patterns, was one of the first to suggest that venation networks might be explained this way. Meinhardt proposed a model that involved two diffusing chemical signals. In this scheme, the signals stimulate branches to grow and divide, while branch tips have a tendency to avoid each other. But growing tips are less strongly repelled by filaments that exist already, and so while tips will not meet end to end, a single tip might intersect and reconnect with an older branch—this is the process of anastomasis that I mentioned above. Thus, Meinhardt's reaction–diffusion model can also generate realistic-looking vascular networks; but it is hypothetical and suffers from the fact that it must postulate two chemical triggers, whereas in plants only auxin has so far been identified as one such.

Steffen Bohn and his colleagues in Paris have recognized a similarity between vascular webs and crack networks in thin layers of brittle material (see page 95). In both cases, for example, branch intersections at right angles are common. They have proposed that, by analogy with cracking, vein networks might be controlled by mechanical forces rather than chemical signals. They point out that leaves contain both soft and stiff layers of tissue, which creates stresses, and they speculate that the directions of vein growth and the angles at which they branch and intersect might be determined by the way veins push and pull on one another.

If nothing else, this illustrates that there is more than one possible way to grow a tree. What seems clear, however, is that nature has found a way of employing some simple principles that achieve a delicate balance of chance and determinism to generate distribution networks beautifully adapted to their role of transporting fluids.

WEB WORLDS

Why We're All in This Together

I n the summer of 2003 the lights went out in New York City. In fact, they went out all over the east side of North America, from Detroit to parts of Canada, affecting one-third of Canada's population and one in seven people in the United States. But there is something about a stricken Manhattan that always plays to the public imagination, and here was the city in unaccustomed darkness, its offices abandoned, its streets benighted, its stock exchange frozen. A state of emergency was declared.

Inevitably, perhaps, many New Yorkers feared at first that this was a terrorist act. But it wasn't. The outage was due to a breakdown of the power grid itself, on a scale rarely witnessed before.

But surely this grid, so vital to the economy, the security and the safety of the country's citizens (imagine if this had happened in the depths of a New York winter), is designed to avoid such catastrophic failure? After all, it is a complex, interconnected web that offers many different routes from one part to another. As in a city road network, if one way is blocked then there is always another. What event could be so damaging and pervasive as to undermine this wealth of alternatives? Culpability was attributed and denied all over the affected territory. The Canadian Department of National Defense blamed a lightning strike in the Niagara region. The Canadian prime minister's office said the cause was a fire at a power plant in New York. The Canadian Defense Minister pinned it to an alleged breakdown at a nuclear power plant in Pennsylvania. The governor of New York State would have none of this, instead placing the blame on the Canadian side of the border.

In early 2004 the final report of a joint US-Canadian task force assigned the job of investigating the largest power failure in American history came up with the following explanation: the power company FirstEnergy had not been diligent enough in trimming trees in part of its Ohio service area. The report said that at 1.30 in the morning on 14 August a generating plant in Eastlake, Ohio, stopped transmitting power because a computer error left it unable to cope with the high electrical demand. This put a strain on nearby high-voltage power lines, and they were tipped over the edge into failure when they came in contact with overgrown trees.

Overgrown trees? Does the north-eastern seaboard of North America stand vulnerable to some tall trees in the woods of Ohio? Was it this that led to the shut-down of 256 power plants by a quarter past four that morning? You might reasonably ask, what kind of idiot designed this system?

The answer is that no one, idiotic or otherwise, designed this system. How could they? A power grid of this magnitude is not built like an office block, according to the blueprint of an architect. It is a growing thing, continually expanding and mutating in response to changes in electricity demand, in demographics, in technology, and the need for repair and renewal. This is not a product of human planning but is, like cities themselves, more akin to an organism, sustained by the complicated interconnections between its component parts.

Civilization generates many complex networks of this sort, which evolve over time according to no master plan. Road networks and urban streets are like this, and so are the webs of trade and travel that link global air and sea ports. The telephone network was one of the first technological artefacts to be considered as a complex network, but it is the emergence of the Internet that has truly placed a spotlight on the interconnectedness of communications. Beyond this, however, there are human networks that are less tangible but arguably even more vital to the ebb and flow of society: our webs of friends and associates, the networks of business and commerce, the virtual conduits along which ideas, money, rumour, culture, and disease are conveyed.

Since none of these systems enjoys much by way of the debatable benefits of rational design, it is quite proper that they be regarded as aspects of what was once called natural philosophy. We do not know what governs their growth, nor what structures result. We must explore them as we would aspects of the natural world. And indeed, it is now becoming clear that they share many characteristics with networks that exist in nature, such as food webs or the communication between genes and proteins in our cells. It is rather remarkable, then, that until the past decade or so scientists have had so little to say about complex networks, their patterns of links and the qualities that these engender. The 2003 North American power failure provides just one among very many reasons why it is a matter of some urgency to understand them, to decode their laws of growth and to map out their structures.

ALL THE WORLD'S A STAGE

Until recently, the only archetypal network one tended to see in human society was that of a tree: the family tree. Today, mapping one's family tree has become a hobby, if not sometimes an obsession, for thousands of people. It is a habit acquired from the nobility, who were traditionally much exercised about the quality of their bloodlines, and also from theologians, who made great play of biblical genealogies, sometimes depicted in medieval art with a very literal allusion to the arboreal analogy (the Tree of Jesse, showing Christ's ancestry, is a favourite theme for stained-glass church windows). There is an obvious conceptual link here with the tree symbolism favoured by early Darwinists, its branches an elaboration of the classical Great Chain of Being.

Now, the striking thing about this network *topos* is that it implies direction and insists on a uniqueness of path: there is only one way to get from the tree trunk to any particular branch tip. To row upstream in a river to the headwaters of any specific tributary, you must select the correct branch every time the channel divides. In genealogical trees, loops are rare—although much less so for the inter-marrying nobility than for typical members of society today (unless they live in a small, self-contained community).

But to understand the structure of our social worlds, a tree is not the right kind of metaphor. If we think instead about our network of friends, we will see straight away that loops are the norm. I know Joe and I know Mary, but of course Joe and Mary know each other, because we all work in the same office. I met my friend Wendy at a party of my friend Dave's, because they are old school pals. So if we map out these friendship networks by drawing links between friends, we find it interconnected in many ways. We might imagine that the more appropriate image, then, is a net or a spider's web. In this network, there are many different ways to get from one intersection (or *node*) to another.

Mathematicians call this kind of network a *graph*, and they study its properties using graph theory. The beginnings of this branch of maths can be traced to the great eighteenth-century Swiss geometer Leonhard Euler, who studied a problem posed by the bridges of the East Prussian city of Königsberg (now Kaliningrad in Russia). There are seven of these bridges across the Pregel river, five of them giving access to an island where the river divides. Can one stroll around the city along a route that takes you over each bridge only once? In 1735 Euler proved that this was impossible. He did so by converting the layout of Königsberg into a graph in which each node represented one of the land areas and the links between them represented the bridges. This was, in effect, the first theorem of graph theory.

Graphs are a form of map, but they generally take no heed of geography, of the spatial distance between points. Instead, they are concerned with the pattern of connectivity, or what is often called the *topology* of the network. Metro maps are somewhat like this: although the spatial positions of the stations on the map are more or less related to their geographical locations, the distances between stations on the map typically have little relation to those in the real world. If you lived your entire life riding the metro, then you wouldn't care about any approximate geographical realism, such as whether a station lay to the north or the east— all that would matter is the connectivity, or how to get from one station to another. And when a graph does not reflect a spatial structure—for example, if it depicts a friendship network—then the actual positions of each node on the diagram have no significance at all; everything of importance is in the topology.

So what might a real friendship network look like? Sociologists and anthropologists have had a long-standing interest in that question. One of the first attempts to give it a thorough mathematical treatment came in the 1950s, when the mathematician Anatol Rapaport at the University of Chicago was seeking to understand how infectious diseases propagate through a population. If the disease is passed on through personal contact, then clearly its rate of transmission depends on how often infected people encounter others who are not infected (or immune). An outbreak in an isolated community might devastate that community but have little impact on the world beyond. By contrast, in a big city where individuals come across many strangers every day, we might expect that the disease would quickly reach epidemic proportions. So there is presumably some critical level of average connectivity, in terms of the number of strangers encountered in a day, above which the outbreak switches from being contained and localized to uncontained and pervasive. At what point does this happen?

Rapaport didn't know what a typical 'encounter network' looked like for any society. But he guessed that it would be reasonable to assume this was pretty random: you have an equal chance of bumping into everyone in the population. Moreover, it seemed sensible to imagine that there was a well-defined average number of such encounters—that is, an average *connectivity*, denoted K. You might, for example, get close enough to 30 other people each day, en route to work or in the shops say, to infect them. Obviously randomness is not the whole story, because there is a much bigger probability that we will encounter (and in this setting, infect) our immediate family members and work colleagues than we will Joe Random. But the assumption of randomness seems a good place to start.

The social network on which the disease spreads, then, is represented here by a series of nodes (people) connected at random to an average of K other nodes (Fig. 6.1a). Rapaport and his colleagues were able to show that in this case the fraction of the total population that becomes infected increases exponentially as the average degree of connectivity K increases: the more connections to each node, the greater the spread of disease. That is what you would expect, of course, but Rapaport put that expectation into precise mathematical terms.

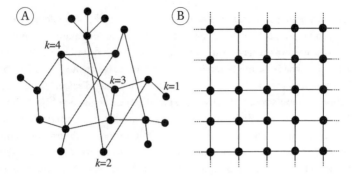

FIG. 6.1 Social networks can be drawn as vertices or nodes (circles), representing individuals, linked by lines that could, for example, represent ties of friendship, or 'infectious' close contacts made during a day, or sexual contacts. There is a distribution of degrees of connectivity k— some individuals are highly connected, others have few connections (a). In a random network or random graph, where links between individuals are forged at random, the distribution of k values has a particular mathematical form, with a well-defined average value. Such networks are disorderly, in contrast to regular grids (b), where all vertices have precisely the same degree of connectivity (in this square grid, the connectivity k is 4).

At the end of the 1950s, the *random graphs* that Rapaport introduced were studied more rigorously by the Hungarian mathematician Paul Erdös and his collaborator Alfred Rényi. They wanted to understand what general properties these graphs have. That is not an easy thing to assess, because by definition there is an immense number of different random graphs that connect any given set of nodes: each graph is constructed by forging links at random between any two nodes, no matter how close or far apart they are. You can 'grow' such a graph by following an algorithm, repeating it again and again:

1. Choose a node at random.

2. Choose another node at random.

3. Draw a link between them.

This procedure does not guarantee that all nodes will have the same number of links; in general, they will not. But each graph has an average number of links per node (K) which increases steadily as you keep iterating the steps above. One of the questions Erdös and Rényi asked was: how well-connected is the graph as K increases? If there are only a few links between many nodes, then some nodes will remain isolated, while others will be connected only into little patches. If there are many more links than nodes—if K is large—then you may have a good chance of finding a route between any two nodes in the network. How does this 'connectedness' of the entire array of nodes alter as K is increased? You might expect it to increase steadily, but Erdös and Rényi calculated that in fact there is an abrupt change at a certain critical value of K equal to 1. This can be seen by looking for the largest interconnected set of nodes in the system. If each node has, on average, less than one connection, this largest interconnected fraction stays small—negligible in comparison to the size of the entire array. But once K exceeds 1, the largest interconnected component grows in size very rapidly, quickly approaching the size of the entire network.

This is critical for problems of epidemiology such as those Rapaport studied. It tells us that, for random networks, the spread of disease can be explosive once the connectivity exceeds a certain threshold, because at that point just about everyone becomes connected to everyone else.

There is another important property of a random graph, which is related to how quick it is to navigate. Suppose you want to go from any one node to any other. If the degree of connectivity is well above the critical threshold, so that essentially all of the nodes are interconnected, there are in general many alternative routes you can take. Measured in terms of the number of links traversed, some of these can be very long. But some are rather short, because the principle of random interconnection forges many shortcuts between nodes that are 'distant' in visual terms. (Recall that in general these visual distances do not have any physical meaning, they're just a way of representing the nodes on a two-dimensional plot. So the 'length' of each link has no meaning either—that is why we can think about path lengths merely in terms of the number of links they incorporate. It is perhaps better to say that the 'shortcuts' ensure that every part of the array of nodes is likely to be accessible from every other part by a few hops, rather than by a succession of many links.)

Just as we can define an average connectivity K for a random graph, so we can define an average path length L: this is the average number of links you must cross to get from one node to another. For a random graph, this length can be surprisingly small even for a big network. We have a familiar expression for this: 'it's a small world'.

It is familiar because that is what experience tells us. You meet someone at a party whom you have never seen before, but it turns out that she went to school with your brother-in-law. In my experience, the older you get, the more this seems to happen. If social networks are random networks, however, it is no surprise. The small-world effect suggests that the random network is a much better description of social networks than a regular grid is, in which nodes are connected only to their immediate neighbours (Fig. 6.1b). In that case there are no shortcuts: to get from one node to a 'distant' one, you have no option but to pass along all of the links in between.

But there is one problem: social networks are *not* random.

This is obvious the moment you stop to think about it, for the reason I alluded to above. What the random-network model says is that two of your good friends have no greater chance of knowing each other than they have of knowing anyone else in the population. But this clearly isn't so. I met Wendy at Dave's party, because Wendy and Dave are old friends. We meet our friends' friends, and they become our friends too. Or perhaps we all met in a group, at work, or at college, or at the sports club. To put it technically, if there are links in a social network from A to B and A to C, then there is a much higher chance of a link between B and C than between either of these and some other, random node. Friends tend to form *clusters* with a strong degree of interconnectivity, while these clusters are linked to other clusters via a lower density of links (Fig. 6.2). The same is true of other social networks: businesses tend to show

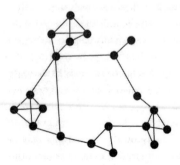

FIG. 6.2 Real social networks tend to be clustered: groups of friends know one another, linking them in a dense web, while there is a lower density of links to other groups.

clustering in their dealings with other companies, as do the collaborations between scientists or musicians.

Clustering seems to work in opposition to the small-world effect of short path length. In effect, clustering implies that there is a higher chance of a node forming a link to another in its vicinity than to one far off: there is a bias against making shortcuts. Indeed, clustering is precisely what one would expect from grid-like graphs, where all links are 'local', joining near neighbours. How can a social network simultaneously show the high clustering characteristic of a grid *and* the short average path length characteristic of a random network?

The physicists Duncan Watts and Steven Strogatz, working at Cornell University in the late 1990s, showed how that can happen. They came up with a model that allowed them to convert a network gradually from a grid-like structure to a random structure. It was called random rewiring, and it was a surprisingly simple idea. You started with a grid, and 'rewired' the nodes one at a time. At each step, a node was selected at random and one of its links was redirected from a near neighbour to a randomly chosen node, which could be near or far. Watts and Strogatz began with a circular grid of nodes, since that has no edges (where the connectivity of the nodes is different from that in the interior). The circular grid on the left of Fig. 6.3*a* may not look like a typical grid, but it is: each node is connected just to its two neighbours and its two next nearest. As the random rewiring proceeds, the regular structure disappears into a tangled jumble of links.

As this happens, both the amount of clustering and the average path length decrease, as expected: these are both large for a regular grid, and small for a random network. But the crucial point is that they don't decrease at the same rate. The path length drops abruptly after only a few rewirings: a handful of shortcuts is enough to bring most of the nodes 'close' to most of the others. But the clustering changes rather little until more rewiring has been effected. So there is a family of networks, for rewiring degrees between these two limits, which have small L but high clustering (Fig. 6.3*b*). These have properties that seem to resemble our social 'small worlds', and Watts and Strogatz called them small-world networks.

Are social networks really like this? It is very difficult to gather data on people's friendship networks—I'll come back to this shortly—but Watts and

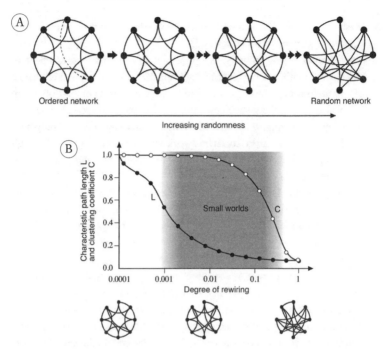

FIG. 6.3 The rewiring networks of Strogatz and Watts are steadily changed from regular grids to random networks by breaking and remaking links at random (a). For a certain range of rewiring between these extremes (0 = regular grid, 1 = fully random network), the networks are 'small worlds', with short average path lengths L between any two points, but significant degrees of clustering C (b).

Strogatz took a different approach to the question. They looked at the network that links film actors according to whether they have appeared together in a film. Each node of this grid is an actor, and there is a direct link between them if they have been co-stars. The virtue of this network is that it is precisely defined—either two actors were in the same film or they were not. Furthermore, the network is already known, thanks to a game invented in the early 1990s by a group of American college students. These film buffs had come to the conclusion that the actor Kevin Bacon was the centre of the film universe. It was not that Bacon was a particularly wonderful actor (though he is fine), nor that he is a particularly big star; but in the movies of that period he seemed to crop up everywhere. He had worked with countless better-known stars. (It helped that his early successes included the ensemble pieces *Animal House* and *Diner*.) In the Kevin Bacon Game, the aim was to link any named actor to Kevin Bacon in as few steps as possible. Elvis Presley? He appeared in *Harum Scarum* (1965) with Suzanne Covington, who appeared with Bacon in *Beauty Shop* (2005). So you can reach Bacon from Elvis in two hops: Elvis has a Bacon Number of 2.

All this can be deduced from the Internet Movie Database, which lists just about all the commercial films made since 1898 (about 200,000 of them). But Watts and Strogatz did not have to comb through it, because the film-star network had already been mapped out by computer scientists Brett Tjaden and Glenn Wasson at the University of Virginia, who had decided to automate the Kevin Bacon Game. At the Oracle of Bacon,* you can find the Bacon Number of any actor in an instant, along with the shortest path linking that actor to Bacon. (That's where I got the Elvis link—I'm afraid I'd never heard of Suzanne Covington.) Tjaden was more than happy to give Watts and Strogatz access to his database.

The researchers hoped that the film actor network would serve as a surrogate for other social networks. Clearly, making a film together is not the same as striking up a friendship (it can notoriously mean quite the opposite), but it seemed reasonable to assume that the two might share some features. And if nothing else, the film-actor net might be a good model of *professional* contact webs. In any event, Watts and Strogatz established that the film network has precisely the characteristic they identified for a small-world network: it has an average path length comparable to that of a random network containing the same number of nodes and links, but much higher clustering.

In fact, the average path length was 3.65, which means that on average any actor can be connected to any other in between three and four hops. Given that the database spans a century and many different countries, that seems remarkable. But what of Kevin Bacon? The average Bacon Number currently stands at about 2.9, which implies that Kevin Bacon is indeed better connected than an average actor—it takes less than three jumps, on average, to reach him. Yet there is nothing so special about that: there are

*See http:// oracleofbacon.org.

many actors with comparably small average path lengths, and over a thousand are better connected than Bacon. The best is currently Rod Steiger: the average Steiger Number is 2.68. As Duncan Watts has pointed out, in a small-world network just about *everyone* seems to be the centre.

The Kevin Bacon Game echoes a familiar trope popularized by John Guare's 1990 play *Six Degrees of Separation*, in which a character claims that 'Everyone on this planet is separated by only six other people.' Where did Guare get that idea from? Why six? In 1967 the Harvard social scientist Stanley Milgram devised an ingenious experiment to measure how well-connected social networks are. He sent 196 letters to randomly selected people in Omaha, Nebraska,** asking them to forward it to a stockbroker from Sharon, Massachusetts who worked in Boston. All Milgram provided was the man's name, along with a curious request: rather than trying to track the stockbroker down, the recipients were to send the letter to someone else they knew personally, who they felt might be better placed to know the man. (Perhaps they might have relatives in Boston, or were stockbrokers themselves, say.) All recipients were asked to do the same, until the letter reached its destination. And surprisingly, some of them did. Even more surprising, however, was the number of journeys those letters took to get there. On average, just six were required: just five intermediaries between the start of the chain in Omaha and the Bostonian stockbroker. A small world indeed.

There are many other indications that our social and professional networks share this property of short average path length. Mathematicians have long enjoyed their own version of the Kevin Bacon Game in which they look for the shortest path that links them with Paul Erdös, the founder of random-graph theory. It is not this distinction that led to Erdös being singled out, however, but the great number of collaborators he worked with on his many papers (he published over 1,500 of them). The average Erdös Number in the network of mathematicians and scientists that can be linked to him this way is about 4.7.* Mark Newman, working at the Santa Fe Institute in New

---

*Mine is 5, so far as I know, but I take heart from the fact that Erwin Schrödinger's is 8.
**Actually they were not all selected at random, and another group of recipients in this study weren't in Nebraska at all—see page 169.

Mexico, has established that the average path length in the collaboration network of the 44,000 scientists who have placed preprints of their papers on the archive set up by physicists at Los Alamos National Laboratory is 5.9: again, six degrees of separation. 'This small-world effect is probably a good sign for science', Newman concludes. 'It shows that scientific information—discoveries, experimental results, theories—will not have far to travel through the network of scientific acquaintance to reach the ears of those who can benefit by them.'

The same small-world property is found in other collaborations. The physicists Pablo Gleiser and Leon Danon in Barcelona have looked at the network formed by over a thousand jazz musicians listed in the Red Hot Jazz Archive who played in bands between 1912 and 1940, and find that it has an average path length of just 2.79.

But Watts and Strogatz also looked at the properties of two networks that have nothing to do with social systems: the electrical power grid of the western United States, and the links between neural cells in the nematode worm *Caenorhabditis elegans*. The latter is one of the best studied multicellular organisms, since it has a relatively small number of cells: precisely 959 in the female and 1,031 in the male. Of these, 302 form a primitive nervous system, in which the pattern of connectivity is known exactly. Watts and Strogatz found that both these networks have the small-world feature of an average path length close to that of the corresponding random network, coupled to a considerably higher degree of clustering.

THE RICH GET RICHER

But what do these networks actually look like? Watts and Strogatz's random rewiring model generated specific examples of small-world networks, but these were rather artificial, being pinned to a ring of nodes in which each node was constrained (by the way the model was set up) to have precisely three links. Clearly, social networks are not like that. But Watts and Strogatz assumed that they are not so different: they imagined that even if we don't all have precisely the same number of friends, this number probably doesn't differ very much. In other words, there will be a certain average number of friends per person, and progressively fewer people will have progressively less or more links than that.

But, Watts confesses with some chagrin, they didn't check that this was really how things are. It turns out that many of the small-world networks in the real world do not look like this at all.

The first intimation of that came in 1999, when Albert-László Barabási and his colleagues Réka Albert and Hawoong Jeong at the University of Notre Dame in Indiana investigated the connectivity statistics for over 300,000 web pages on the domain of their university. Each web page is connected to others by hyperlinks, where a click on the mouse will take you to the other page. The researchers found that this segment of the World Wide Web (WWW) is a small world: they estimated that, if it is representative of the Web as a whole, then any web page is connected to any other by an average of just 19 links—even though the WWW at that time held over a billion documents. And because of this small-world structure, that average path length will not increase much even at the rapid pace with which the WWW is growing. Even if the number of pages expands tenfold, the average distance between pages would grow by just two links. This slow growth in average path length as a network expands is one of the characteristics of a small-world network, and is the result of the large number of shortcuts that very quickly weave new nodes into the web.

But the pattern of links was quite different from what Watts and Strogatz had implicitly assumed. Most pages had only a single link, though some had hundreds, and a very few had several thousand.* Now, you might intuitively expect there to be many more pages with rather few connections than with many; but the connectivity statistics that Barabási and colleagues found are different from what you might expect if the connections were being made independently and at random. In fact, they have a very particular mathematical form: the number of pages having k connections is proportional to the inverse of k raised to some power. This is, in other words, yet another example of a power law. We saw in Chapter 2 that systems whose structure is described by power laws have the general property that they are *scale-free*—there is no characteristic size

---

*The picture is slightly complicated by the fact that there is a difference between incoming and outgoing links. A hyperlink only takes you in one direction—you cannot necessarily get back to your starting page from the one you end up at via a hyperlink. But Barabási and colleagues found the same statistical pattern for both incoming and outgoing links.

to them. What this means here is that there is no 'typical' number of links to a node of the network.

These connectivity statistics tell us what the probability is of a node picked at random having a particular number of connections k. Although this chance gets smaller as k gets bigger, it does so considerably less quickly than it would either for the kind of networks obtained from random rewiring or for random graphs. In other words, the WWW has a disproportionate number of very highly connected pages. This means that, when the network is plotted out on a page, parts of it seem to be 'pinched' where a large number of links converge on these highly connected nodes (Fig. 6.4a). In contrast, a random graph looks relatively uniform throughout (Fig. 6.4b). The physical structure of the Internet (by which I mean the actual links between computers, not the virtual net that connects web

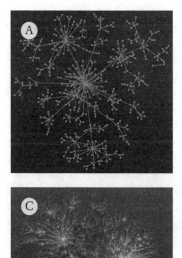

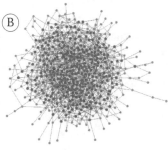

FIG. 6.4 Scale-free networks look 'pinched' at a few highly connected vertices (a), whereas random graphs are rather uniform (b). This pinched quality is evident in the structure of a part of the Internet (c; see http://www.cybergeography.org/atlas/topology.html). (Images: a was prepared using NetLogo software, available at http://ccl.northwestern.edu/netlogo/; b: courtesy of Paros Oikonomou and Philippe Cluzel, University of Chicago.)

pages) also has this shape (Fig. 6.4c).* It means that not all of the nodes are equal—some enjoy much better connections than others. There are many nonentities, but a few celebrities.

It now seems that many diverse networks have this same topological structure, with power-law distributions of connectivity. The pattern is found, for example, in email communications (where a link is established between two nodes if an email is sent between them), in the web of direct flights between airports, in the network of trade links between countries— and in the network linking movie actors. Outside the human social sphere, Barabási and his co-workers have found scale-free networks in the bio-chemical pathways of living cells. For example, they looked at how all the molecules involved in the metabolic chemical reactions of *Escherichia coli* bacteria, and the protein enzymes in brewer's yeast, interact with one another, with a link existing between them if they participate together in a particular chemical process. In both cases the network was scale-free (Fig. 6.5).

Where does this topological structure come from? Graph theory has traditionally considered the different ways that a bunch of nodes may be wired together; in the random graphs of Erdös and Rényi, links are made between two randomly selected nodes. But Barabási and Albert realized that many real networks *grow* almost like plants branching from a seedling: they approached the origin of scale-free networks as a problem of growth and form. The World Wide Web is growing every day as new web sites are created and new pages added. Whenever a new node is plugged into the network, the question is, to which others should it connect? A rule that the link is always made to the nearest node, for example, would tend to generate a grid. A rule that the selection is made at random will produce a random graph. But Barabási and Albert said that scale-free networks grow according to another rule: the new node connects to an existing node at random, but with a *bias*: the more links a node already has, the more likely

*An analysis of this structure—actually that of a subset of over 4,000 nodes of the Internet— was conducted by the brothers Michalis, Petros and Christos Faloutsos, all of them computer scientists, in 1999, at much the same time as Barabási and colleagues were mapping the WWW. The Faloutsos brothers found precisely the same kind of power-law relation for the connectivity of nodes.

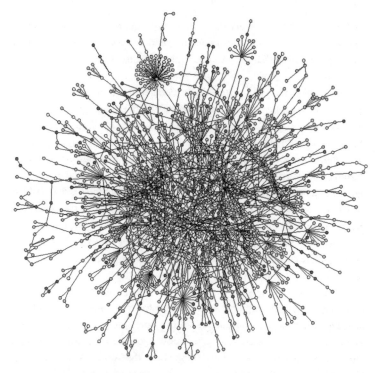

FIG. 6.5 A part of the network formed from molecules involved in yeast metabolism. Each vertex is a molecule, and the links denote enzymatic reactions that convert one molecule to another. (Image: Hawoong Jeong, Korea Advanced Institute of Science and Technology.)

it is to be chosen. A node with two links is twice as likely to be selected as a node with one.

This means that the more connections a node already has, the more likely it is to acquire more as the network grows. But that doesn't mean that all the best-connected nodes are *guaranteed* to be awarded all the new links, because there is an element of chance in the choice. If there are lots of nodes, then even an extremely well-endowed one will have only a relatively small chance of being selected compared with the chance that a new link will go instead to any one of the others. That is why the most well-connected nodes in a scale-free network are also the most rare. All the same, the implication of this rule of 'preferential attachment' is plain to see: in terms of connectivity of nodes, the rich get richer. The connection rule guarantees a persistent inequality of connectedness—and Barabási and Albert showed that the nature of this inequality is described by power-law statistics.

In retrospect, this is no surprise. The 'rich get richer' principle is precisely what seems to happen in capitalist societies: wealth attracts yet more wealth. If that is so, we should expect it to lead to a power-law distribution of wealth, in which a very few individuals are obscenely rich. In 1897 the Italian economist and sociologist Vilfredo Pareto showed that this is the case in many societies: at least for the rich end of the income distribution, the figures follow a power law, which economists now call the Pareto law. The American sociologist Robert Merton has dubbed this the Matthew Principle, since the Gospel of Matthew provides one of the first known descriptions of this particular injustice: 'For unto every one that hath shall be given, and he shall have in abundance; but from him that hath not shall be taken away even that which he hath.'

Why, though, should networks grow according to this principle? There isn't really a general argument that works for every scale-free net from metabolic pathways to film actors, but in social networks it often boils down to the issue of fame. The better known you are, the more likely your fame will be boosted even more. Think about the World Wide Web: you have made a new web page and want to provide a link to some standard reference source for a particular aspect of what it describes. The chances are that you will choose the link that you can see others have chosen for the same purpose. Of course, these days you're likely to find that link by a Google search, but that makes the bias even more strongly deterministic, because the Google page ranking scheme depends on how many links a page has.* And so pages acquire links not because someone has reviewed all the alternatives and decided that it is the best reference source, but because it is already 'famous'. The same is true for citations in the scientific literature, which also have power-law ranking statistics: people cite a book

---

*The page rank is a little more sophisticated than that, since it also takes into account the connectedness of the pages from which the links are coming; but the Matthew Principle still operates.

or a paper because that is what others have done, and not because they have read it themselves.

This doesn't mean that a node's connectedness in a network like this bears no relation to any genuine merit. It may be that a web page or a citation begins to attract more links than others because it really is good. But as long as the law of preferential attachment operates, the connectedness of different nodes probably does not reflect their real differences in merit, because the effect of fame artificially inflates some nodes over others, perhaps in ways that even invert the real distinctions of quality between them.

WORLD OF WEBS

Does this mean that all small worlds are formed from scale-free networks? Not at all. After all, Watts and Strogatz began the whole story by constructing small worlds that did not have this property. And it turns out that power grids are not scale-free either: the electricity grid of southern California, for instance, doesn't show the characteristic power-law relation between the connectivity of its nodes. That might have something to do with the fact that power grids are physical entities that exist geographically in two dimensions: this makes it very hard to establish extremely highly connected nodes, because any node can only have a limited number of near neighbours, and direct links from very distant nodes are uncommon. The same is true of road networks, which are highly interconnected but not scale-free, and indeed are not even small-world networks at all: they are more like regular grids.

Even scale-free networks generally have a limit: in theory there is no ceiling to the number of links a node might have, but in practice there is. For example, no actor, however much in demand, can make a million films before they retire or die. The capacity of an airport is ultimately limited by the number of runways, facilities, and so on. What this means is that the statistics deviate from a power-law relationship for very high connectivities so that the probabilities are lower than the power law predicts. Luís Amaral at Boston University and his colleagues showed in 2002 that this is true even for the WWW network, simply because the web is too vast for anyone choosing to make a hyperlink to survey the entire system before selecting the target: the information about the web has to be filtered before it can be processed at all, which means that only a subset of all the available nodes is ever considered.

What about real social networks of friends: are they scale-free? This isn't clear, because it is extremely hard to gather data, and harder still to know how general it is. A study of an acquaintance network among 43 Mormons in Utah, conducted in 1988, and another of 417 secondary-school students in Wisconsin in the 1960s, both seem to show a statistical distribution that is not scale-free but has a well defined average number of links. Nonetheless, these networks are small worlds in the sense that any one person is linked to any other by only a small number of links.

There is another important characteristic of scale-free networks that gets rather lost if we focus only on the power-law scaling of the connectivity of nodes. If you look at the graphical representation of the Internet in Fig. 6.4c, one of the things that strikes you first is that it seems to be built up from a number of clusters: densely radiating groups of nodes that look like the head of a dandelion in seed, linked to one another by a more sparse web of links. There is, in other words, a number of distinct *communities* within this web. On the Internet, these modules seem to be derived largely from geography: each module corresponds to the sub-network of an individual country. There are also sub-modules that reflect particular professional communities, such as military sites. It is no surprise that a community structure exists in many if not most social networks, for that is after all a reflection of how our lives tend to be organized. In the network of scientific collaborations, those people working in the same discipline, and in the same sub-field of a discipline, are likely to be bound into a community. Friendship networks might be structured around a neighbourhood or a workplace. In some sense, modularity is a reflection of the high degree of clustering characteristic of small worlds.

And yet it is possible to create scale-free networks that do not have this modular community structure. So the fact that some of them *do* is telling us that there is more to the shape of a network than is revealed simply by its connectivity statistics. But it is not always an easy matter to find this community structure: to work out what the modules are and where their boundaries lie. One of the problems is that, because there is a strong element of randomness in the way these networks grow, two groups of nodes might have only a small number of links between them purely by chance, rather than because they genuinely constitute distinct communities in a meaningful sense. The task is then to work out not just if two groups are joined by few links, but if they are joined by fewer links than we would expect purely by chance.

Various techniques have been devised for teasing community structure out of complex networks. Extracting this 'buried' information is often of great value, helping to make sense of what otherwise might look like just a mess of 'wiring'. In a metabolic network, for example, it could tell us something about how a cell's biochemistry is organized into functional modules. Mark Newman has used a community-finding scheme to reveal the undercurrents in purchases of books on US politics through the online bookseller Amazon.com. He studied a network of 105 recent books, representing nodes, which were linked if the Amazon site indicated that one book was often bought by those who purchased the other. The analysis showed a clean split into communities containing only the 'liberal' books and only the 'conservative' ones, as well as two small groups that contained a mixture along with some 'centrist' titles. Newman found a similar political split in links between over a thousand blogs. This clear division, Newman says, 'is perhaps testament not only to the widely noted polarization of the current political landscape in the United States but also to the cohesion of the two factions'; put another way, it suggests that people only want to read things that reinforce their own views.

Newman has highlighted another aspect of the deep substructure of complex networks. In some of them, highly connected nodes show a greater-than-average propensity to have links with other highly connected nodes, forming what has become dubbed a 'rich club'. This phenomenon is known as assortative mixing. Obviously, the existence of rich clubs in social, economic, and professional networks could have an enormous impact on the way society functions—it might imply, for example, that the 'rich club' members are able to share privileged information that percolates only slowly into the rest of the network. Newman has shown that while neither the random graphs of Erdös and Rényi nor the scale-free networks grown from Barabási and Albert's 'preferential attachment' model has rich clubs, many real-world social networks do, including the collaboration networks of scientists, film stars, and company directors. On the other hand, some natural networks are negatively assortative: they show fewer than expected links between 'rich' nodes. This is true of the Internet, and to a small degree of the WWW; and it also applies to the protein interaction network of yeast, the neural network of the nematode worm, and the marine food web. In other words, there is something different about social networks compared to others, either technological or biological: we humans seem disposed towards forming rich clubs. Newman has argued that this is because of the strong tendency of social networks to partition into communities, which is less prevalent in other webs.

SEARCHING THE WEB

One can begin to see, then, that the shape and form of networks can have a crucial bearing on how it performs its function. For instance, how does the topology affect the ease with which a network can be navigated? Clearly, we might expect to be able to get around rather quickly on any network that has the small-world property of a short average path length between nodes, because we can always find a shortcut. But that could depend on having a map, for otherwise how do we know which is the best link to follow? Search engines are of course invented with the explicit aim of conducting that search for us, but even the best of them cannot always intuit our wishes from a handful of search criteria, and in any case the WWW is now so vast that no search engine can index and encompass it all.

The key characteristic of social networks revealed by Milgram's epochal experiment is not simply that they are small worlds but that they are *searchable* small worlds. It would be of no value that a short social path exists between two individuals if we were unable to find it. But some-how—and this is what is truly surprising—we do, at least some of the time. Why is that?

Actually, there is a lot that is swept under the carpet in this 'some of the time'. The truth is that the rate of 'successful searches' in a case like this is rather low. One fact about Milgram's experiment that is often overlooked is that the starting points in his chains were not by any means all randomly selected citizens of Omaha, Nebraska. Of the roughly 300, about 100 were stock investors (and remember that the target was a Boston stockbroker), and another 100 lived not in Nebraska but in Boston! And of the 96 letters that began their journey from Nebraskans picked genuinely at random, only 18 reached the destination. In other words, there was a big drop-out rate. Some of this can be attributed to mere apathy. But Duncan Watts and his colleagues have found that whether a journey like this is completed seems also to depend on the perceptions of the participants about their proximity to the target. In 2003 they staged a re-run of the Milgram experiment using email, which made it easier to enlist volunteers: they accumulated more than 60,000 in 166 different countries. And the targets were similarly varied: 18 in total, in 13 different countries, with a wide variety of backgrounds. Of the 24,163 chains that were started, only 384 were completed, again showing that this kind of search has a high attrition rate. But one particular target had a much lower drop-out rate than the others: an American Ivy League professor. The total path lengths for these chains were not significantly greater than the others; it was just that they did not break halfway. Since over half of the participants were college-educated Americans, Watts and colleagues figured that these chains were more likely to run to the end because the intermediaries believed that they would: they thought it more worthwhile forwarding a message intended for an American academic than for an Estonian archivist, because it was more likely to succeed.

All the same, it remains surprising that the small-world nature of social networks gets exploited at all in these experiments, since there is no way for any intermediary to know that he or she is really routing the message in an efficient direction. In 2002 Duncan Watts, Mark Newman and the mathematician Peter Dodds set out to understand why this happens. They suggested that the community structure of the network, and in particular the identification of its members with specific groups, is an essential part of the searchability. They supposed that each individual mentally breaks down the network into hierarchical arrangements of groups. They do this in more than one way, for example imagining groups based on profession and on geographical location: these different hierarchical pictures overlap. This gives each individual a picture of their 'social distance' from others. Most members of the network lie way over this social horizon and are not even considered: people think only about their local links to those a short 'social distance' away. The researchers explored a wide range of ways in which individuals might conceptualize the hierarchy of the network and forge links between neighbours who are more or less similar to themselves. They found that in most cases the resulting networks were searchable in the sense that, if individuals route a message only using their local view of the network, its average path length between origin and destination is small.

In other words, people do not need to see the whole map; a parochial view is enough to ensure efficient routing, as long as everyone feels that they are part of overlapping communities. It is this multiplicity of communities that is the key to the social small world, the researchers claimed, because that is what allows 'unexpected' shortcuts to be found. Joe might pass a message to Mary because they work in the same office; but then Mary passes it to her brother Ray, who lives in another country. But Joe doesn't even know that Ray exists—he has never spoken to Mary about her family. So long as Mary didn't come into the conversation, Joe and Ray would say that the social distance between them is huge. The shortcut via Mary passes through two different 'social dimensions'—work and family—and so is not apparent to either Joe or Ray in the absence of other information, as it would be if, say, Joe, Mary and Ray were all working in the same profession.

This picture of a 'multidimensional' social web is supported by research by mathematician David Liben-Nowell of Carleton College in Minnesota and his co-workers. They studied the network formed by around half a million bloggers in the online community LiveJournal who live in the United States. The attraction of this network is that members explicitly list other members who they consider to be friends. This allowed Liben-Nowell and colleagues to deduce how much friendship depended on geographical proximity. For an online community, one might expect that geography hardly matters, but in fact the researchers found that two-thirds of the friendships depend on the physical distance between the individuals. This of course means that the remaining one-third of the friendships are independent of geography, and arise because of some perceived similarity in some other 'social dimension'.

The researchers then ran another Milgram-style email-routing experiment on this network. They didn't actually get the bloggers to do the forwarding, but instead conducted computer simulations of what would happen if each person were to send the message to the friend who was geographically nearest to the target. In this way, 13 per cent of the messages reached the target in just over four steps. This seemed puzzling, because the computer scientist Jon Kleinberg had shown that routing a message according to geographical proximity along a grid does not allow the network to be navigated at all efficiently—there is no small-world effect.

The reason for that apparent contradiction, said Liben-Nowell and colleagues, is that the chance of becoming friends does not simply decrease the further apart they are, but depends on how many other people are nearer: in other words, it depends on the size of the pool from which geographically closer friends might be selected. The chance of two people becoming friends depends on how many others lie in between—which is a question both of geographical distance and of the density of others in that gap. A network like this is quite different from the uniform geographical grid that Kleinberg considered, and has the virtue of small-world navigability.

When you want to pass something like information around on a network, a small-world nature is beneficial: it allows the distribution to reach all corners of the net pretty efficiently. But there are some things that we would rather *not* see spreading through our social and techno-logical networks—for example, diseases and computer viruses. Biologists have studied the transmission of disease through populations for over a century, but it is only recently that they have started to realize how important a role the topology of the network might play in this. The standard model for studying epidemics supposes that individuals come in two classes: healthy and infected. If a healthy person encounters an infected one, he or she has a chance of becoming infected. But meanwhile, infected individuals can recover and become healthy again. The disease then spreads at a rate that depends on the relative probabilities of infection and recovery. This so-called susceptible-infected-susceptible (SIS) model says that if this spreading rate exceeds some threshold value, the disease becomes an epidemic, sweeping through the entire population and per-sistently infecting some constant proportion. But if the rate is smaller than this threshold, the disease dies out. The aim, then, is to keep the spreading rate below the epidemic threshold, for example by vaccinations that make the infection probability small enough.

The physicists Romualdo Pastor-Satorras and Alessandro Vespignani looked at how the SIS model plays out if the encounters between people are described by a scale-free network. This changes the behaviour significantly: there is no longer an epidemic threshold, and all diseases can pervade the network no matter how slow-spreading they are. That fits with what we find for computer viruses, which are transmitted by messages passing through the scale-free email network: they are frustratingly hard to eradicate completely, remaining active in the system by persistently infecting a very small fraction of computers.

How about human diseases? The way they spread depends on whether encounters between people create a scale-free network. In the age of cheap air travel, there are surely plenty of shortcuts between distant populations, which is one of the big concerns about the threat posed by new infectious diseases, such as novel and virulent strains of influenza that threaten to emerge from the avian flu virus H5N1. In 2005 Dirk Brockmann at the Max Planck Institute for Dynamics and Self-Organization in Göttingen, Germany, and his colleagues found some evidence that our freedom to travel might create a scale-free small world of human contacts. They used data collected by automated systems for tracking bank notes by their serial number to look at how nearly half a million dollar notes 'travelled' around the contiguous United States, and found that the relationship between the distance travelled and the time taken for these notes followed a power law, just as it would if the notes were being passed around on a scale-free network. If most of these notes move from place to place in people's wallets, they give a fair indication of how humans (or at least, Americans) get about. All the same, this is not by any means a direct indication that humans encounter and infect one another in a scale-free manner.

Some encounters are more intimate than a brushing of elbows on the train, and may give you rather more than a cold. AIDS is now the third biggest cause of premature death in the world, and kills two million people a year in sub-Saharan Africa alone. Networks of sexual contacts are the fatal grid on which HIV spread, and it is hard to develop effective strategies for attacking the epidemic unless we know something about the patterns it traces out. If this, too, is a scale-free net, it would seem that prospects for eradicating the virus entirely are dim. We simply do not know, however, if this is the case or not. A study of 3,000 Swedes in 2001 suggested that the distribution in the number of sexual partners over a twelve-month period followed a power-law; but the result remains controversial.

COMMUNICATION BREAKDOWN

The implication of these findings is that flows—of information, rumour, and disease, say—on scale-free networks are not easily disrupted. For human health, that sounds like a disheartening message. But for techno-logical networks like the Internet, this aspect of scale-free topologies would appear to be a virtue. It implies that, if a few links get broken, the network is in no danger of falling apart. The multitude of pathways privileged with small-world shortcuts means that you can always find a good alternative for routing your emails. That idea has been confirmed by Albert, Jeong, and Barabási, who have compared the way scale-free and other networks (such as random graphs) fragment as more and more nodes are 'killed' (all links to and from them being severed) at random. In random graphs these 'connection failures' soon make the network break apart into isolated clusters, so that it becomes impossible to reach any point in the web from any other (Fig. 6.6a). But scale-free networks break up differently: there remains a large cluster of interconnected nodes even for rather severe amounts of link failure. This central cluster merely sheds little 'islands' of nodes as the damage gets worse (Fig. 6.6b). The network doesn't shatter, but gently deflates.

Since breakdowns do happen—servers get jammed or malfunction—this is surely just the kind of property one would like in a network like the Internet. But no one designed it that way. Indeed, if they *had* designed it, they probably would have chosen some other network topology that was nothing like as resilient. The Internet acquired this happy feature simply from the way that it grew.

Yet scale-free robustness has a cost—one might call it an Achilles' heel. The resilience of the network depends on the presence of a few highly connected hubs: the richest of the rich, which offer shortcuts between

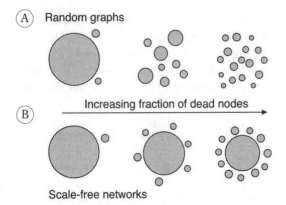

FIG. 6.6 How webs fall apart. As increasing numbers of nodes are deactivated, so that the links to them are effectively severed, at random, a network breaks up into isolated units. But this happens much more quickly for random graphs (*a*) than for scale-free networks (*b*): the latter tend to 'deflate', shedding small islands but retaining a large, interconnected core.

many different regions. Now, if one were to deactivate nodes not at random but in a targeted fashion, taking out only the most highly connected hubs, the story is very different. While the resilience of the Internet to random breakdowns is extraordinary—it has been estimated that a connected cluster of nodes that reaches right across the network remains even for almost 100 per cent breakdown of links—it becomes highly vulnerable to an attack that knocks out the most highly connected nodes first. With a few well-placed shots, you could scupper the entire network. Albert, Jeong, and Barabási have calculated that if such a strategy deactivates just 18 per cent of the nodes, the Internet would be shattered into many tiny pieces. That is now one of the concerns of organizations worldwide that have been established to combat the threat of cyberwarfare: intentional disruption of computer networks. With ever more aspects of our lives being dependent on these information pipelines, from health services to power supplies, this is emerging as a serious concern.

There is a positive side to this vulnerability of scale-free networks. Since it seems they exist in the biochemical pathways of living cells, describing the interactions of protein enzymes for example, then targeting drugs at the 'hub sites' of pathogenic organisms or rogue cells might be a good way of killing them off. And while diseases may spread faster and become harder to eradicate in scale-free social networks, immunization and vaccination programmes aimed at the key hubs—the most highly connected individuals, for example those who are most sexually active—may have an inordinately positive impact. Barabási and his colleague Zoltán Dezsö have shown that treating the hubs against a viral infection in fact restores the threshold for the virus to spread as an epidemic, making it possible to eradicate it entirely. Of course, in the real world it may be far from easy to identify who the hubs are, or to reach and treat them. But an immunization strategy that does even a rather crude job of finding and treating highly connected nodes preferentially can reintroduce an epidemic threshold to a scale-free network, making it easier to contain the virus so that it may die out naturally.

What, finally, might these studies of network structure have to tell us about power failures such as the one that turned out the lights of Manhattan? It is not clear whether any power grids are scale-free networks—possibly some are and some are not. But in any event, they *do* in general appear to be small worlds, with many shortcuts and small average path lengths between nodes. And this seems to confer the same kind of mixture of robustness and vulnerability. Random failures usually do not matter, because alternative routes can be found for the electricity. But such networks do seem prone to a particular kind of catastrophic breakdown in which a few local failures create cascades: overloads get passed on down the line quickly, escalating as they go. This seems to be what happened in the 2003 US blackout, and very probably in the even larger one a month later that affected the whole of Italy. A local failure means that the electrical load gets passed to another part of the grid, which in turn becomes overloaded and shuts down, and so the problem gets shunted further down the line, leaving failed links in its wake. It is possible to design networks that do not have this tendency for cascading breakdown—they are generally *not* small worlds, but require rather long average path lengths between nodes— but for technological networks that grow without any central planning, such as power grids and computer networks, design is not really an option.

Yet cascade failures may not be inevitable for small-world networks, so long as they are understood within the context of the network topology in which they occur. Once the pattern of the network is taken into account, it might be possible to tailor an appropriate response strategy. Dirk Helbing at the Dresden University of Technology and his co-workers have proposed that in such cases the best strategy is to reinforce the most highly connected nodes first against failure. That makes intuitive sense and is what you might have guessed anyway—but only once you appreciate the way the network is configured in the first place. You first have to see the pattern you are dealing with.

EPILOGUE

THE THREADS OF THE TAPESTRY

Principles of Pattern

Nature uses only the longest threads to weave her patterns, so each small piece of her fabric reveals the organization of the entire tapestry.

Richard Feynman, *The Character of Physical Law*

Nature is an endless combination and repetition of a very few laws. She hums the old well-known air through innumerable variations.

Ralph Waldo Emerson, 'History', *Essays*

It is time to take stock. I hope it has become apparent in the course of these books that there is going to be no Grand Theory of Pattern awaiting you at the rainbow's end. The universe is not made that way. Some physicists have in recent years encouraged an unfortunate aspiration towards grand unified pictures, but the world that we encounter, the world of real stuff that we see and touch, is far too messy for such things.

But this does not mean we must capitulate to a welter of detail. I have shown that there are fundamental patterning processes that recur in nature, operating identically in many different settings. Laplacian growth instabilities lead to snowflakes, or soot, or cracks in pavements and continents. Convection orders clouds, stones, and pans of hot milk. Reaction–diffusion processes may generate the leopard's spots and the graveyards of ants. What nature uses is not a Law of Pattern but a palette of principles. And there is, I submit, much more wonder in a world that weaves its own tapestry using countless elegant and subtle variations, combinations and modifications of a handful of common processes, than one in which the details become irrelevant and in which a few recondite equations are supposed to explain everything. To my mind, the astonishing thing about many natural patterns is not just that we can implicate a few basic processes that lie at the heart of them all, but that small changes in the details, in the specific initial or *boundary* conditions, produce such fantastic variety—think, for example, of butterfly wings. By the same token, the patterns of a river network and of a retinal nerve are both the same and utterly different. It is not enough to call them both fractal, or even to calculate a fractal dimension. To explain a river network fully, we must take into account the complicated realities of sediment transport, of changing meteorological conditions, of the specific vagaries of the underlying bedrock geology—things that have nothing to do with nerve cells.

The late Rolf Landauer, an uncommonly perceptive and broad-thinking physicist, has expressed very pithily this need to resist over-enthusiastic aspirations to the universal: 'A complex system is exactly that; there are many things going on simultaneously. If you search carefully, you can find your favorite toy: fractals, chaos, self-organized criticality, Lotka–Volterra predator–prey oscillations, etc., in some corner, in a relatively well developed and isolated way. But do not expect any single simple insight to explain it all.'

Perhaps it is a shame to begin a summary chapter with a caution against too much summarizing, but that is for the best. For the ideas that form the backbone of our understanding of spontaneous pattern formation seem so powerful, so all-encompassing, that they are all too often paraded as the keys to a theory of everything. Even D'Arcy Thompson would not have wanted us to believe that.

And yet what is extraordinary and thrilling is that so many pattern-forming systems have so much in common, to the extent that by understanding one we can predict a great deal about the others. This realization has made a delicious mockery of the traditional, rigid divisions between scientific disciplines, so that physicist, economist, ecologist, chemical engineer, and geologist can all talk to one another—*and in the same language*. When this happens, something very exciting is going on in science.

Yet we have seen that many of the ideas behind pattern formation are not new. Oscillating chemical reactions were known in 1901; convection cells showed up around 1900; and Kepler sensed the reason for the sixfold symmetry of a snowflake in the seventeenth century. But D'Arcy Thompson was unable to persuade most of his peers of the importance of form and pattern in the 1920s, and only in the past two decades or so has pattern formation emerged as anything like an identifiable field of study in its own right. Why is that?

One reason is the explosion in computer power. Many of the theoretical ideas about patterning are tough to test experimentally, since there are so many factors to bring simultaneously under control; but as we have noted again and again, computers allow researchers to perform 'ideal' experiments in which everything can be repeated exactly and complicating factors included or excluded at will. Many theoretical models are simple in principle but utterly impossible to test by manual number-crunching—the calculations would take for ever if done by hand. But although this increase in computational capacity has provided scientists with perhaps the most important technological tool currently at their disposal, it also serves to underline the phenomenal achievements of early researchers such as Michael Faraday, Lord Kelvin, and Lord Rayleigh, Geoffrey Taylor, and Andrei Kolmogorov, who had to rely on their exquisite intuition alone to deduce the essential physics of pattern-forming processes.

I believe there is another reason, little emphasized but equally important, for the recent development of ideas in pattern formation. This is the maturation since the mid-1980s of a field of theoretical physics that provides much of the framework for understanding the features that frequently characterize spontaneous patterning, such as abrupt, global changes of state and scaling laws. The field is the study of phase transitions and critical phenomena, and it is the largely unseen bedrock of all physics today. I shall say more about this discipline in what follows.

Let me now try to pull together some of the threads that have run, more or less perceptibly, through all the previous chapters. They do not, even collectively, constitute a 'theory of patterns'. Rather, these ideas are like stepping stones that lead us through the turbulent ebb and flow of pattern and form in the physical and natural world.

COMPETING FORCES

Spontaneous patterns typically represent a compromise between forces that impose conflicting demands. The ordered architectures adopted by certain polymers and soap-like molecules (Fig. 7.1: see Book I, Chapter 2) are an elegant solution to the requirement of keeping the structure's surface area and curvature small while also packing molecules together efficiently. The expanding waves and the static spot and stripe patterns of some chemical mixtures (Fig. 7.2: see Book I, Chapters 3, 4) result from a delicate balance between reaction and diffusion of the molecular components, and between short-ranged amplification and long-ranged inhibition of their chemical reactions. The bulbous pseudopodia of viscous fingering (this book, Chapter 2) are the manifestation of competition between an

FIG. 7.1   Ordered structures in co-polymers: a balance between surface area and curvature of the interfaces of domains and the efficient packing of molecules. (Photo: Edwin Thomas, Massachusetts Institute of Technology.)

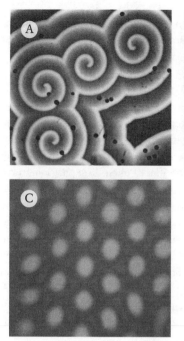

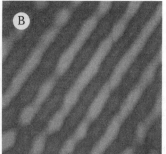

FIG. 7.2 Spirals, stripes, and spots in oscillating chemical reactions. (Photos: a Stefan Müller, University of Magdeburg; b, c, Harry Swinney, University of Texas at Austin.)

instability that generates branches at the interface, and surface tension, which limits its size scale.

Snowflakes are the product of a similar compromise forged in an environment that is imprinted with an intrinsic microscopic symmetry. Trains of breaking-wave vortices appear in fluid flows when an instability that excites waves wins out over the tendency of viscosity to suppress them (Book II, Chapter 3)—a victory from which a preferred patterning size emerges.

Competition lies at the heart of the beauty and complexity of natural pattern formation. If the competition is too one-sided, all form disappears, and one gets either unstructured, shifting randomness, or featureless homogeneity—bland in either event. Patterns live on the edge, in a fertile borderland between these extremes where small changes can have large effects. This is, I suppose, what we are to infer from the clichéd phrase 'the edge of chaos'. Pattern appears when competing forces banish uniformity but cannot quite induce chaos. It sounds like a dangerous place to be, but it is where we have always lived.

SYMMETRY BREAKING

When spontaneous patterns appear in systems that are initially uniform, this lowers or *breaks* the original symmetry. That is why it is important not to confuse symmetry and pattern: high symmetry does not by any means imply the richest patterns, and indeed often those that look most striking have rather low symmetry, or none at all. On the other hand, symmetry tends to break a little at a time, and we may find that the least symmetrical patterns are those that appear when the driving force responsible for them is greatest, so that the system is furthest away from an equilibrium state. I shall say something shortly about *why* symmetry breaks away from equilibrium. At this point, we need appreciate only that symmetry breaking is not like laying a tiled floor. A hexagonal cellular arrangement like that of Rayleigh–Bénard convection (Fig. 7.3: Book II, Chapter 3) is not a matter of imposing cells of an

FIG. 7.3 A hexagonal array of convection cells in Rayleigh–Bénard convection. (Image: David Cannell, University of California at Santa Barbara.)

arbitrary shape, one by one, in an inert medium. Rather, the medium becomes everywhere at once imbued with a 'hexagon-forming tendency', once the driving force (here the heating rate) exceeds the pattern-forming threshold. It then takes only the smallest fluctuation to release this 'hexagon-ness' globally. The hexagonal array seems to rise out of the floor, you might say.

NON-EQUILIBRIUM

Nearly all the pattern-forming systems in these books are out of equilibrium—that is to say, they are not in their thermodynamically most favourable state. Once scientists considered such systems to be unapproachable, perhaps even unseemly. Thermodynamics, the science of change that developed initially as an engineering discipline in the nineteenth century, was intended to describe the equilibrium state of systems. It told one about the direction of change, and allowed one to calculate the amount of useful work that could be extracted from that change; but what actually took place *during* a change was something that classical thermodynamics could barely touch. It was a pretty good tool for chemical and mechanical engineers who wanted to gauge the performance of their machines. But it offered a rather artificial view of the world in which everything happens in a series of jumps between stable states that do not otherwise alter over time. That is not very like the world we know. Thermodynamics was silent in the face of the uncomfortable fact that some processes *never* seem to reach equilibrium. A river does not simply empty itself into the sea in one glorious, ephemeral rush—the water is cycled back into the sky and redeposited in the highlands for another journey. And so will it always be, while the sun still shines.

Out of this somewhat restrictive picture, however, emerged the idea of an *arrow of time*. Nearly all processes seem to have a preferred direction: they go one way but not the reverse. Heat flows from hot to cold, an ink droplet disperses in water. These processes are said to be *irreversible*. One-way processes fit with our intuition—ink droplets do not re-form—but they become somewhat puzzling when we look closely at the microscopic events behind them. In the mathematical equations that describe how a particle of ink pigment moves in water, there is no arrow of time: you could play a film of the particle's motion backwards and not notice the difference, nor appear to break any physical laws. It is only when you look at the behaviour of the whole ensemble of particles that you would notice anything odd when time is reversed and the droplet coalesces from a glass of uniformly dilute ink.

Most scientists agree that irreversibility has its roots in the second law of thermodynamics, which states that in a system isolated from its surroundings (so that it cannot exchange energy or matter), the direction of change is always towards greater *entropy*, which loosely means, towards greater disorder. This is a law enforced not by physics but by probability. It emerges when a system has many choices of how to arrange its components: there are simply many more disordered arrangements than ordered ones. We encountered the second law in Book I, where I explained that it prompted objections to the idea of oscillating chemical reaction in the 1950s. As entropy is in some sense a measure of disorder, the second law seems to pose a big problem for the spontaneous appearance of pattern.

Thermodynamics allows one to predict what a system at equilibrium will look like: it tells us that such systems adopt states for which the total energy is as small as it can be.* This is the *energy minimization* criterion that we saw in operation in Book I, controlling the shapes of soap films. Is there an analogous criterion that determines what the state of a non-equilibrium system is?

The thermodynamics of non-equilibrium systems is concerned not with some end point in which entropy has increased in relation to the initial state; rather, it considers the process of *becoming*, of how change occurs. Since the second law specifies that an irreversible change leaves a system with more entropy than it began with, change *produces* entropy. When the system reaches its new equilibrium, entropy production ceases. So the process of change seems to be bound up with the issue of entropy production.

---

*You may recall from Book I that what is really being minimized here is a particular quantity called the free energy, or the Gibbs energy.

In the 1930s the Norwegian-born scientist Lars Onsager began to explore that connection. He considered change driven by only a small departure from equilibrium. Under these conditions Onsager deduced universal laws relating the driving forces to the rates of entropy production. For this work he was awarded the Nobel Prize for chemistry in 1963. The Russian-born chemist Ilya Prigogine, working in Brussels, added an important element to the picture by showing that, in this near-equilibrium case, non-equilibrium systems tend to behave in a way that *minimizes* the rate of entropy production. If prevented from reaching equilibrium (where the rate of entropy production is zero), the system will instead settle into a steady but dynamic state in which entropy is produced at the lowest possible rate. Here, then, was a criterion for determining the preferred state of a system out of (but close to) equilibrium.

Nothing in this prescription, however, gives any hint that away from equilibrium a system may stumble into a *patterned* state with orderly structure, like Bénard's convection cells or Turing's spots. States like this tend to emerge rather far from equilibrium, where the equations formulated by Onsager and others no longer apply and Prigogine's minimum entropy production rule can break down. Where do they come from?

During the 1950s and 1960s, Prigogine and his colleague Paul Glansdorff attempted to extend the treatment of non-equilibrium thermodynamics to the more interesting far-from-equilibrium situation. They were able to show that, as the force driving a system away from equilibrium increases, the steady state of minimal entropy production reaches some crisis point where it breaks down and becomes transformed to another state. Technically speaking, there is a *bifurcation*—literally, a branching in two—at which the evolution of the steady state splits into two branches, presenting a choice of new states that the system can adopt (Fig. 7.4).

What are these new states? The theory of Prigogine and Glansdorff said little about that; but it seemed reasonable to suppose that they might correspond to the self-organized structures and patterns that were known to appear far from equilibrium. To progress any further, however, we first need to appreciate how these regular or ordered non-equilibrium states are fundamentally different from superficially similar ordered states in equilibrium systems.

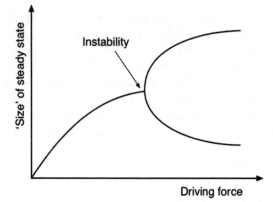

FIG. 7.4 A bifurcation occurs when a stable state develops an instability that offers the system a choice of two new states. A 'pitchfork' bifurcation like that shown here commonly occurs as a system is driven further out of equilibrium.

DISSIPATIVE STRUCTURES

Regularity is not rare. A swinging pendulum, a bouncing ball, the 'green-grocer's stall' atomic packing of a crystal, the yearly passage of the Earth around the Sun: all are periodic in space or time. But the regular hex-agonal lattice of Rayleigh–Bénard convection differs from the hexagonal atomic lattice of a crystal like copper metal. The latter is an equilibrium structure whose periodicity is determined by some characteristic dimen-sion of the components—here, the sizes of the atoms. The former, meanwhile, is maintained away from equilibrium by a throughflow of energy, which it dissipates in the process, thereby generating entropy. Stop the input of energy—let the top-to-bottom temperature gradient equalize—and the pattern goes away. Likewise, the oscillations of the Belousov–Zhabotinsky (BZ) reaction (Book I, Chapter 3) persist only while the reaction is fed with fresh reagents and the products are re-moved. Patterns that are supported away from equilibrium by the gener-ation of entropy are called *dissipative structures*.

In contrast to most equilibrium structures, the spatial scale of the pattern features in a dissipative structure bears no relation to the size of its constituents. The size of convection cells is much, much larger than the size of the circulating molecules, for example. Furthermore, this size scale persists in the face of perturbations. A convecting fluid may lose its pattern temporarily if stirred, but once the disturbance has passed, the same structure reforms. The system 'remembers' how it is supposed to look. These robust dissipative structures are said to be governed by an *attractor*: a place to which they are 'anchored' in the abstract space of parameters describing the system. An example of such an attractor is the limit cycle of the oscillating BZ reaction. The opposite of a dissipative structure is a *conservative* structure, which possesses no attractors and so can be altered arbitrarily. An example is the orbit of a planet around the Sun: if the radius of the orbit is altered (say, by a catastrophic collision with another body), it stays that way rather than returning to its former value.

If a dissipative structure is 'kicked' too hard, however, it may find itself closer to a different attractor—a quite different state or pattern—and end up being drawn towards that instead. Attractors are like dips in a hilly landscape: if you get pushed over a crest, you fall down into a different basin. The point here, then, is that non-equilibrium systems tend to change not gradually but in discrete jumps between different dissipative attractor states. Let's look at this in more detail.

INSTABILITIES, THRESHOLDS, AND BIFURCATIONS

Most of the patterns that I have described appear suddenly. One moment there is nothing; then you turn the dial of the driving force up a notch, and everything is abruptly different. Stripes appear, or dunes, or pulsations. This seems to be the nature of most symmetry-breaking processes: they happen all at once. In that respect, they resemble *phase transitions* in equilibrium thermodynamics.

Phase transitions are generally abrupt jumps from one equilibrium state of matter to another: from ice to water, water to vapour, magnet to non-magnet. These are 'all-over' transformations. When water cools through its freezing point, we don't find part of it turning to ice and the rest remaining liquid.* Below zero degrees centigrade all the water is ready to

---

*In practice we see that combination quite a lot—a layer of ice on a pond, say. But this is because the water may not all be below freezing point, or because it takes time for the water to freeze or the ice to melt. In those situations, the pond isn't in thermodynamic equilibrium.

become ice, and will surely do so given enough time. And this is an all-or-nothing affair: the temperature need be only a fraction above freezing point for all the ice to have melted once equilibrium is reached. A fraction below, and all is frozen.

In other words, there is a threshold that, once crossed, leaves the entire system prone to a change in state. Just the same is true for many pattern-forming processes. Convective patterns, as we observed above, appear above a threshold heating rate, and vortices in fluid flow above a threshold flow rate. The path of a crack goes crazy above a particular crack speed.

In addition, the change in state during an equilibrium phase transition may involve a breaking of symmetry. Crystalline ice has an ordered molecular structure (in fact it has many ordered structures), while liquid water is disorderly at the molecular scale. Again, you could be forgiven for thinking that symmetry is therefore broken during melting, but in fact it is the other way around: symmetry is broken during freezing, because whereas the liquid state is isotropic (all directions in space are equivalent) the crystal structure of ice identifies certain directions as 'special'.

Thus equilibrium phase transitions, like the abrupt transitions that characterize much of pattern formation, are spontaneous, global, and often symmetry-breaking changes of state that happen when a threshold is crossed.

Some of these transitions involve a straightforward rearrangement of one state into another. But there are classes of both equilibrium and non-equilibrium transitions that offer a choice of *two* alternatives for the new state, which are equivalent but not identical. Think of the formation of convection roll cells. Adjacent rolls turn over in opposite directions, but any particular roll could rotate either one way or the other as long as all the others switch direction too. Above the convection threshold, there is a choice of two mirror-image states. Which is selected? Clearly, there is nothing to favour one over the other, and the issue is decided by pure chance. The same is true of the rotation of plughole whirlpools, unless some small outside influence tips the balance.

The equilibrium behaviour of a magnetic material like iron displays a comparable choice. In iron's magnetized state, all the atoms act like little bar magnets with their north and south poles aligned. If you heat a magnetized piece of iron above 770 degrees centigrade (its so-called Curie point) this alignment is lost, because it is overwhelmed by the random, jiggling effect of heat. The magnetic fields from each atom then cancel out on average, and the piece of iron as a whole is no longer magnetic. This abrupt change at the Curie point from a magnet to a non-magnet is an example of a phase transition. It might seem that this phase transition involves a single choice: either the iron is magnetic or not. But in fact there are two possibilities as the metal is *cooled* from a non-magnetized, randomly oriented state through the Curie point: the atomic magnetic poles can all point either in one direction or the other (Fig. 7.5a). The states are entirely equivalent,* and again the choice depends on random fluctuations that tip the balance. (You may wonder how, or if, this random choice can be made the same way throughout the entire system. I'll come back to this.) The situation is like a ball perched on top of a perfectly symmetrical hill (Fig. 7.5b): it is unstable at the top and has to roll down one side or the other, but which way it goes is unpredictable and at the mercy of imperceptible disturbances.

Freezing and melting of water might seem rather similar to this magnetic transition, in the sense that they too involve atomic-scale order being overwhelmed by, or recovered from, thermal randomness. But there is an important difference, somewhat technical, but important. Freezing and melting are said to be *first-order* phase transitions. Among the distinguishing features of these are the fact that the switch in state begins at one or more randomly selected points and spreads from there throughout the whole system. Freezing starts from a little 'seed' or nucleus of ice somewhere in the water. And there is a step-like change in key properties of the system: in this case, in the density (water is denser than ice). And it is possible for the *less stable* state to persist beyond the transition threshold in a precarious state that is said to be metastable, and which is liable to switch at any time. Water can be *supercooled* below freezing point without turning to ice, if it is free from small particles on which ice crystals might nucleate. This means that in practice the transition might happen

---

*In reality the Earth's magnetic field could supply a bias in favour of one orientation. Indeed, this is how changes in the state of the geomagnetic field are deduced from the imprint it leaves on magnetic rocks that cooled from a molten state many millennia ago.

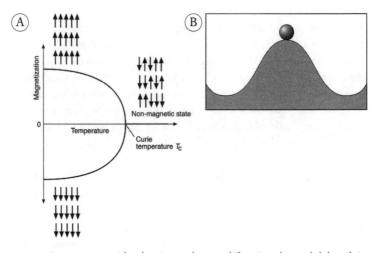

FIG. 7.5 A magnetic material such as iron undergoes a bifurcation when cooled through its so-called Curie point (at temperature T_c), where the material becomes spontaneously magnetic (a). This happens because the atomic magnetic poles, represented here as arrows, all line up in the same direction; above T_c that is prevented by the randomizing influence of heat. Below T_c the magnetization can point in one of two equivalent directions. This kind of bifurcation is analogous to the situation of a ball perched on top of a hill between two identical valleys (b); it must roll down one way or another, but the choice is arbitrary.

at a different point from where equilibrium thermodynamics says it should and, specifically, the threshold might be different as the system passes through the transition in one direction (freezing) to the other (melting). This is called *hysteresis*. Finally, first-order phase transitions may but do not have to involve symmetry-breaking.

The spontaneous magnetization of iron at the Curie point, on the other hand, is an example of a *second-order* or *critical* phase transition. The magnetization changes abruptly but *continuously* as the system goes through the transition—there is no sudden jump from one value to another (see Fig. 7.5a). Second-order and other critical phase transitions *always* involve symmetry breaking. And there can be no hysteresis: the switch to a new state cannot be delayed. It doesn't depend on the formation and growth of some nucleus of the new state, but happens through a kind of global convulsion, driven by the random fluctuations of the components.

FIG. 7.6 Convection patterns in a fluid may switch from targets to spirals in an abrupt transition. But because this transition is of a technical type called sub-critical, targets and spirals can coexist. (Photo: Michel Assenheimer, Weizmann Institute of Science, Rehovot.)

Now, many of the pattern-forming bifurcations that I have discussed are analogous to critical phase transitions—they are called *supercritical* bifurcations, and they lead to symmetry breaking. The onset of convection is like this, as is the switch between hexagonal and striped chemical Turing patterns (Fig. 7.2). But a few pattern-forming processes are *subcritical* bifurcations, analogous to first-order phase transitions, such as the switch between spiral and target patterns in convecting fluids (Fig. 7.6: see Book II, Chapter 3). This is why both spirals and targets can coexist in the same pattern. These distinctions may seem rather esoteric; but I'll show shortly that the details of what happens at a critical phase transition provide important clues to how patterns arise away from equilibrium.

THE SOLUTIONS REVEAL MORE THAN THE EQUATIONS

These analogies with phase transitions are useful, but they don't provide any kind of rigorous mathematical description of pattern-forming bifurcations. Physicists like analogies, but they prefer rigour. Attempts to develop a more concrete description of what happens at a symmetry-breaking bifurcation in non-equilibrium systems began in earnest in 1916 when Lord Rayeigh looked for a theory that would explain Henri Bénard's convection patterns. In the 1920s, Geoffrey Taylor attempted much the same for the case of Taylor–Couette flow between rotating cylinders (Book II, Chapter 6). As I explained in that volume, a thorough treatment of any problem in fluid flow must start with the Navier–Stokes equation—Newton's law of motion applied to fluids, which relates the changes in fluid velocity at every point to the forces that act on the fluid. Both Rayleigh and Taylor looked for the solution to the Navier–Stokes equation appropriate to their respective systems *close to equilibrium*—that is, when the driving force for patterning was very small. They then examined how this 'base state' would respond as the system was driven increasingly far from equilibrium.

For the case of convection, the base state is one in which no flow at all occurs: if the temperature difference between the top and bottom of the fluid is small enough, heat flows simply by *conduction* through the fluid, just as it does when it spreads through a solid. The question is then, 'does the base state remains stable if it is disturbed a little?' Rayleigh considered an all but imperceptible wavy disturbance in the fluid, with a particular wavelength. Below the critical Rayleigh number Ra_c (a measure of the bottom-to-top temperature difference) at which convection begins, such perturbations die away over time regardless of the wavelength, and the system returns to the base state. But exactly at Ra_c, something stirs: one particular wavelength neither decays nor grows, although all others decay. And just above Ra_c the 'special' wavy perturbation grows: a pattern with this wavelength develops. As Ra increases, the range of possible wavelengths of the convection pattern broadens steadily. Above Ra_c the base state is still a solution to the Navier–Stokes equation for the system—but it is an *unstable* solution, since the slightest disturbance will trigger the appearance of a pattern at one of the allowed wavelengths.

The key point about this approach, which is called linear stability analysis, is that it can be applied to a wide range of systems that undergo spontaneous patterning—not just those in fluids that obey the Navier–Stokes equation. A similar kind of analysis can be carried out, for example, for the equations that describe reaction–diffusion systems. Some researchers have attempted to construct simplified 'model equations' that describe the generic features of pattern-forming systems such as these while avoiding the complexity (and thus the intractability) of, say, the full

Navier–Stokes equation. They are general-purpose equations for predicting patterns arising from an instability in an unpatterned base state, regardless of the detailed nature of the particular system in question.

Perhaps the most important message to have emerged from these studies of instabilities is that we do not necessarily have a complete understanding of a system once we know the equations that govern it. What we really want to know are the *particular solutions* to those equations. The latter need not be obvious from the former. That is a point worth remembering in all of mathematical science. The English physicist Freeman Dyson says that 'to discover the right equations was all that mattered' to Albert Einstein and Robert Oppenheimer in their later years. One might say the same about some physicists working today to develop a 'theory of everything'. But if you take this view, then fluid dynamics was solved once we could write down the Navier–Stokes equation. Yet if we had stopped there, we would never have guessed at the rich variety of solutions that it held in store even for relatively simple experimental situations. In fact, sometimes even knowing the mathematical solutions is not enough: we need to do experiments to interpret what they are telling us.

PATTERN SELECTION

Linear stability analysis can reveal the point at which a non-equilibrium system is driven across the threshold of pattern formation. But can it tell us anything about the pattern that results? Exactly at the threshold, there is (at least for the cases considered by Rayleigh and Taylor) a single 'marginally unstable' wavelength that defines the characteristic pattern scale. But how does this length scale manifest itself—as stripes, spots, or perhaps travelling waves? And once the threshold is surpassed, an increasing range of patterning wavelengths may become stable. Which wavelength is selected?

This question is not peculiar to these examples in fluid dynamics. Just about any system with pattern-forming potential faces choices: it may have a gallery of designs from which to select. Soap molecules on the water surface (Book I, Chapter 2), metal deposits grown at electrodes (this book, Chapter 2), bacterial colonies (Book I, Chapter 5 and this book,

Chapter 2), jumping sand grains (Book II, Chapter 4)—they all confront a catalogue of riches. Which to choose?

There is no universal way to answer this question. Yet there is in this respect a major distinction between equilibrium and non-equilibrium systems. The former may also have several choices of pattern and form, but there is in that case a simple rule (in principle) for deciding the best choice: as we have seen, at equilibrium a system will always seek to adopt the configuration that has the lowest free energy. This means that balls roll down hills, iron rusts in air, water below freezing point turns to ice. A few of the complex patterns I have discussed do form under equilibrium conditions: for example, the shapes of soap bubbles and the self-organization of hybrid polymers (Book I, Chapter 2). If we know the various contributing factors to the free energy, we can predict the selected pattern in these cases by finding the shape that minimizes it.

What about non-equilibrium systems? Might there be some analogous 'minimization principle' for them? I shall return to this question shortly.

We can first make a few general observations about how patterns form away from equilibrium. This normally involves symmetry breaking; and symmetry tends to break in stages, a little at a time, as the system is driven harder and harder. This alone enables us to understand why two types of pattern, stripes and hexagons, are particularly common. The simplest way to break the symmetry of a uniform, 'flat' (two-dimensional) system such as a shallow layer of fluid—that is, the way to break as little symmetry as possible—is to impose a periodic, wavy variation in just one direction (Fig. 7.7a), which produces parallel bands, stripes, or rolls. Parallel to the stripes, symmetry is not broken: as we travel through the medium in this direction, we see no change in its character. It is only in the perpendicular direction that we can identify the broken symmetry: travelling in this direction, we see a periodic change from one state to another and back again. Thus, stripe-like patterns are often the first to appear from a uniform, flat system. That is what we see for sand ripples and in the appearance of convection and Taylor–Couette roll cells.

After breaking symmetry periodically in one dimension, the next 'minimal' pattern in a two-dimensional system involves breaking it in the other, dividing up the system into compartments or grids. If the state is to remain ordered and as symmetric as possible, there are only two

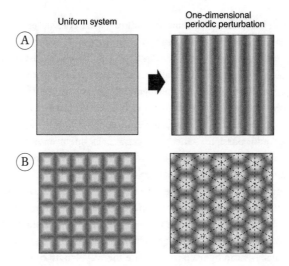

FIG. 7.7 Breaking symmetry by degrees. The simplest way to break the symmetry of a uniform two-dimensional system is to impose periodic variations in one direction, creating stripes (a). When symmetry is broken in both dimensions, square or triangular cells result (b). In the latter case, the resulting pattern can be a triangular or a hexagonal lattice of cells.

options: to impose the periodic variation perpendicular to the rolls, creating square cells, or to impose two such variations at 60° angles, creating triangles or hexagons (Fig. 7.7b). So the square, triangular and hexagonal patterns that we have seen in Turing patterns (Book I, Chapter 4), in convection (Book II, Chapter 3) and in shaken sand (Book II, Chapter 4) are no mystery. They arise simply because the *geometric* properties of space constrain the ways in which symmetry can be broken.

These hand-waving arguments are not, however, a reliable guide for determining exactly what kind of broken symmetry will arise in any particular case. For that, one needs to look into the messy specifics. For example, the question of whether stripes/rolls or hexagons will be preferred has no universal answer. And a linear stability analysis of convection will not help you to answer that: you have to use a more sophisticated level of theory to discover that rolls are generally favoured. Qualitatively we can understand this preference on the grounds that rolls do not distinguish between upflow and downflow (what goes up on one side of the cell is mirrored by what goes down on the other), whereas hexagonal cells do: upflow occurs in their centre and downflow at the edges. So rolls are geometrically symmetrical about the plane midway between the top and bottom of the fluid, but hexagonal cells are not. If this midplane symmetry is broken, however—for example, if the warmer fluid near the bottom is significantly less viscous than the cooler fluid towards the top (which is quite possible)—then hexagonal cells might appear instead. In the case of the Turing instability of reaction–diffusion systems, on the other hand, hexagonal spots are usually the default option.

The other aspect of pattern selection for these relatively simple cases is the *size* of the features: the wavelength of the stripes or the separation of the hexagonal spots. Again, it is not possible to identify a single criterion that determines this. For Rayleigh–Bénard convection, we saw that linear stability analysis can be used to calculate the wavelength of the instability at the onset of patterning. But above that threshold there is a range of allowed wavelengths, and then the width of the roll cells becomes dependent on the *history* of the system: how it reached the convecting state. This size may in fact vary throughout the system, or may change over time. In a chemical Turing system, meanwhile, the scale of the pattern is set by how fast the ingredients diffuse (recall that one component acts as an activator to initiate the formation of a pattern element, and the other as an inhibitor that suppresses other elements nearby). And we have seen that these reaction–diffusion systems may generate moving patterns (travelling waves) rather than stationary ones. That's what happens if, above the patterning threshold, a wavy disturbance to the system doesn't just grow in strength but also itself moves.

As we have noted, abrupt transitions between different patterns commonly occur at threshold values of the driving force. There is a common tendency (particularly clear for fluid flow) for the patterns to become more ornate—we might say more complex—as the system is driven harder and harder. This sequence of increasing complexity is also evident in the oscillating Belousov–Zhabotinsky reaction conducted in a continuous-flow stirred-tank reactor (Book I, Chapter 3). As the flow rate of chemicals through the vessel increases, the oscillations undergo a series of period-doubling bifurcations so that the cycle repeats with every

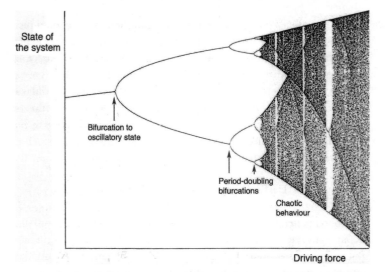

State of
the system

Bifurcation to
oscillatory state

Period-doubling
bifurcations

Chaotic
behaviour

Driving force

FIG. 7.8 Bifurcations in many non-equilibrium systems come in a sequential cascade as the system is driven further from equilibrium (here from left to right). At each bifurcation, the number of states of the system doubles. In an oscillating system such as the Belousov–Zhabotinsky reaction, each bifurcation corresponds to a period-doubling: it takes two, then four, then eight cycles for the system to return to a given state. This cascade structure gets increasingly finely branched and eventually gives way to non-periodic, chaotic behaviour, seen here as a dense 'dust' of dots.

oscillation, then with every second oscillation, then with every fourth and so on. This can be depicted as a *cascade* of bifurcations (Fig. 7.8). One might liken this very crudely to the excitation of additional harmonics as a trumpeter blows harder. Eventually the oscillations become chaotic, as though the system becomes overwhelmed with options. Then the cascade loses its branched structure and breaks up into a dense forest of spots—and we lose sight of any order at all.

THE RACE FOR DOMINANCE

When patterns are *growing*, pattern selection may depend on how, or how rapidly, the different candidates advance. We saw in Chapter 1 that the characteristic scale of branches on a dendritic crystal is set by the condition that the growth speed just about balances the tendency for the tip to split:

there is a unique, 'marginally stable' pattern where growth only just outpaces a propensity to split repeatedly. In other cases, for example non-equilibrium electrodeposition or the growth of a bacterial colony, it has been proposed that the pattern selected is simply that which grows fastest and which therefore outruns the others. But it is not clear that this is a criterion that applies to branching growth in general. Pattern selection is also influenced by noise—by which I mean the inevitable randomness in the environment, such as that supplied by thermal fluctuations. Noise does not necessarily affect all patterns equally; it may favour some over others.

DEFECTS AND BOUNDARIES

Very often patterns that appear far above the initial instability threshold are far from symmetrical; they are laced through with 'mistakes', sometimes to such an extent that all appearance of symmetry is lost. For example, we saw how the roll cells of convection or the stripes of Turing structures can merge, and how the hexagonal cells of Rayleigh–Bénard convection can become grossly imperfect honeycombs. Through an accumulation of such distortions, parallel stripes can become bent into more or less disordered wavy patterns, and a hexagonal lattice of spots can disintegrate into a jumble. This gives us something akin to the stripes of zebras and the spots of the leopard. In some patterns of this sort, perfect regularity is only ever a kind of Platonic dream—the Giant's Causeway merely hints at its relation to the honeycomb. Notice, however, that even in cases where disorder overwhelms all semblance of symmetry, we can still identify order of a kind: the average distance between spots or stripes, or the average number of sides of a polygonal pattern element, remains more or less constant.

Defects have their own logic and taxonomy, enabling us to 'decode the mess' by considering how generic defect structures arise from characteristic deformations of the underlying pattern (Book II, Chapter 3). In attempting this, scientists may often draw on a rich existing theory of defect formation developed from studies of crystals and related materials, such as liquid crystals.

The principles I have adduced so far apply to 'infinite' systems, by which I mean ones for which we ignore the boundaries. But of course no real pattern-forming system is infinite—they always have edges.* If the size of the system is vastly greater than that of the pattern's characteristic length scale, the effects of edges may be negligible except close to the edges themselves. Commonly, though, this is not the case: the pattern may be influenced throughout by the size or shape of the 'container'. We saw in Book I, for example, how either hoops or spots could be selected from the same pattern-forming mechanism on animal tails, depending on the size and shape of the embryonic tail when the pattern is laid down during development. And, more generally, we saw how the patterns of different animal pelts—a two-tone division of the whole body, say, or a few large blotches or a multitude of small spots—can be determined by the relative size of the embryo at the patterning stage. On ladybird wings, both the size and the curvature of the surfaces may affect the patterns.

The shape of a boundary can occasionally change a pattern to something qualitatively different from how it would look in an 'infinite' container. In long, rectangular trays, convection rolls tend to form stripes, whereas in circular dishes the rolls curl up into concentric circles (Fig. 7.9a,

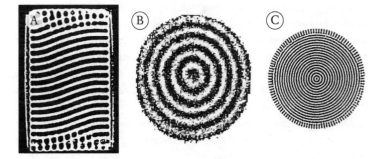

FIG. 7.9 Coping with boundaries. In these convection patterns, the shape of the vessels has imposed particular shapes on the global arrangements of the roll-like cells. (Images: a, from Cross and Hohenberg, 1993, after LeGal, 1986; b, David Cannell, University of California at Santa Barbara; c, from Cross and Hohenberg, 1993.)

*That is not strictly true: there is a wealth of interesting work, for example, on the patterns that form on spheres and other self-enclosing surfaces, such as toruses. Although these do not experience edge effects, the patterns are still constrained by the overall size of the surfaces.

b). Moreover, the need for a whole number of pattern features to fit within the container may determine the wavelength, just as the wavelength and thus the frequency of an organ note is determined by the length of the pipe. In some systems, the pattern may change locally to adapt to the presence of a boundary—in Fig. 7.9*c*, for example, concentric roll cells give way to short parallel rolls at the edges, so that the rolls can meet the boundary at right angles (which is a more stable configuration).

CORRELATIONS AND CRITICAL POINTS

Many of these self-made patterns seem to acquire a miraculous ability to measure and mark out space. It is one thing to talk mathematically about a 'marginally stable' wavelength, but how can we understand this in terms of the interactions between the fundamental components of the system? The scale of sand dunes, for example, is so far removed from the size of the grains themselves, or of the hops they make when landing on a surface, that it is hard to imagine where the pattern scale really comes from. How do the grains 'know' where to start and stop piling up into a new dune? Just the same is true for Turing patterns: the size of the molecules and atoms in the chemical mixture, and the range of the interactions between them, is minuscule (about a tenth of a millionth of a millimetre), yet the patterns have length scales big enough for us to see with our unaided eyes, perhaps several millimetres or so. How on earth can interactions on these unimaginably tiny scales give rise to patterns millions of times larger?

The implication seems to be that the components of the system are able to 'communicate' with each other over distances much longer than those to which they are accustomed at equilibrium. Think of the rolls that appear in Rayleigh–Bénard convection. Before the onset of convection, the molecules are moving about throughout the quiescent fluid in a random, disorderly way; each molecule barely takes heed of what its immediate neighbours are doing, let alone what is happening a millimetre or so away, many millions of molecules distant. Yet above the patterning threshold this independence has been lost, and the molecular motions have (on average) become *correlated* over these vast distances. That is to say, if we were to observe the molecular motions on the descending edge of one of the roll cells, we would know that statistically identical motions were being executed by molecules one wavelength away—and two, and three, and so forth throughout the vessel. This kind of long-ranged correlation, according to which molecules behave coherently over distances that far outstrip the sphere of their own influence, is characteristic of many pattern-forming systems.

How is it possible? Are the molecules able to relay their individual, tiny influences from neighbour to neighbour over such scales? That is quite out of the question: in the frenzied environment of a hot liquid, it is like trying to play Chinese whispers at a rock concert.

The appearance of long-ranged correlations in systems undergoing abrupt changes in behaviour is not unique to non-equilibrium systems. It has been long recognized in equilibrium phase transitions, too. The key to such behaviour, both at equilibrium and away from it, is that the system loses all sense of scale. Long-ranged correlations may develop when a system becomes *scale-invariant*: the correlations exist over every range.

This is what happens in the critical phase transition of iron at its magnetic Curie point, which is an example of a *critical point*. I explained in Book II (Chapter 4) that, at a critical point, fluctuations occur on all length scales. I showed that fluids also have critical points, where the distinction between gas and vapour states vanishes. We saw that the distribution of gas and liquid regions at this critical point is scale-invariant: the domains occur at all scales (Fig. 7.10). This picture could equally well show domains in a piece of iron where the magnetization points in one direction or the other (see Fig. 7.5). As the iron is cooled down through its Curie point, the magnetic poles of all the atoms switch from being randomly oriented to being aligned. But right at this point, there is no telling which of the two orientations (shown as black or white here) will prevail. The critical system is infinitely sensitive to fluctuations: the slightest imbalance suffices to tip it one way or the other.

As a system like this approaches its critical point, each component feels the influence of more and more of the others. Far from the critical point, only the behaviour of a magnetic atom's nearest neighbours influence its own alignment. But as the critical point is approached, each atom's sphere of influence (called its correlation length) extends wider and

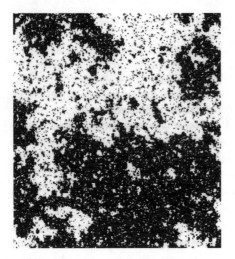

FIG. 7.10 The scale-invariance of domain sizes at a critical point, such as the Curie point of a magnet. This is a fractal structure. (Image: Alastair Bruce, University of Edinburgh.)

wider. And exactly at the critical point, the correlation length becomes 'infinite'—which is to say, as big as the entire system. This does not imply that the magnetic field set up by each atomic magnet becomes stronger, or reaches out further; rather, the behaviour of the atoms becomes more *collective*, so that progressively larger groups will behave cooperatively.

In this regard, non-equilibrium pattern formation can resemble a critical phase transition: it relies on long-ranged correlations between components to allow them to become organized at large scales. But there are important differences, particularly in terms of the structures that result. For in our piece of iron, the ordering that results as we pass through the phase transition has a characteristic length scale that reflects the scale of interatomic forces: the magnetized iron adopts a regular structure in which the periodicity of the magnetic alignment occurs on the same scale as the periodicity of the atoms. We could have guessed this length scale even above the critical point, for it is essentially the same as the range of each atom's magnetic influence. In non-equilibrium patterns this is not so: the scale of ordering vastly exceeds the range of interaction of the constituents, and there is no obvious hint of a length scale of this magnitude in the microscopic physics of the unpatterned state. This is why we can regard such patterns as global *emergent* properties of the system, which are likely to remain hidden to a reductionistic analysis of the microscopic interactions.

POWER LAWS AND SCALING

Most of the discussion so far has been concerned with patterns that form in systems that are essentially *deterministic*, which is to say that at least in principle we can write down equations (such as the Navier–Stokes equation) that describe the behaviour exactly. That does not by any means imply that we can solve the equations, but it follows that, once the initial and boundary conditions (the rate of heating, say, and the size of the container) are specified, we know what all the ingredients of the process are.

Some of the patterns that I have talked about in these books, and particularly in this final volume, do not share this deterministic character. The equations that describe them contain a strong random element that is unpredictable and impossible to formulate in anything other than statistical, average terms. Diffusion-limited aggregation (Chapter 2) is like this: the particles become attached to the growing DLA aggregate only after a random walk through space, jostled by air molecules say, so that their precise trajectories are indeterminate in advance. Noise is an essential ingredient of the patterning process in these cases, and the forms that result look rather disorderly.

The characteristic, invariant forms of systems like this are commonly 'hidden'—they are mathematical rather than visual. We saw in the case of sand-pile avalanches (Book II, Chapter 4) how the apparently random sizes of slope collapses are governed by the 'hidden regularity' characteristic of self-organized criticality, distinguished by power-law statistics. And the most robust feature of disordered fractals like DLA clusters or city shapes is their power-law scaling behaviour or fractal dimension, which enables us to classify and compare structures that bear no *particular* visible features in common. Some researchers believe that power laws are the key to understanding much of the complex behaviour exhibited by non-equilibrium systems that are strongly influenced by noise. But it remains unclear to what extent self-organized criticality offers a general framework for describing such systems. Nonetheless, it is clear that noise, power-law behaviour, scale invariance, avalanche behaviour and fractal forms are intimately connected in some deep way that remains to be fully explored and unravelled.

THE ROLE OF ENTROPY PRODUCTION

No doubt this all looks like a rather piecemeal approach to the issue of pattern selection in non-equilibrium systems. During the 1960s and 1970s, Ilya Prigogine's group at Brussels held out the hope of finding a more general criterion: a 'minimization principle' analogous to the minimization of free energy at equilibrium. In other words, the selected pattern in each case would be one that minimizes some quantity. As we saw earlier, Prigogine showed that systems only slightly out of equilibrium observed the principle of minimum entropy production. But this principle did not seem to hold in general for systems further from equilibrium, which is where spontaneous patterns form. It now appears that there is probably no such minimization principle that can be applied in general to all non-equilibrium systems. To address the problem of pattern selection, we are forced to consider the specific details of each system. In many cases the only option is to resort to experiment, to characterize and categorize the taxonomy of dissipative structures that appear.

Yet there *is* at least one candidate for a 'general' principle of non-equilibrium pattern selection that applies very widely, if not universally. It stems from work begun in the 1950s by the American mathematical physicist Edwin Thompson Jaynes, who attempted to reformulate the microscopic, molecular-scale foundations of thermodynamics. In the late nineteenth century, James Clerk Maxwell, Ludwig Boltzmann, and others showed how the laws of thermodynamics governing the properties of heat and matter could be understood by considering the behaviour of individual atoms and molecules as they jiggle and collide. This discipline became known as statistical mechanics, since it drew on the average behaviours in the molecular melée. Boltzmann showed what entropy really means at this microscopic scale: it is a measure of the different ways molecules can be arranged. In the 1940s an American engineer named Claude Shannon showed that the concept of entropy could be applied not just to molecules but to *information*: it could quantify the ways in which units of information may be arranged. Jaynes then attempted to unite Shannon's 'information theory' with the statistical mechanics of Boltzmann and his successors. This enabled him to start developing a kind of statistical mechanics that applied also to non-equilibrium systems too—something that Lars Onsager had imagined but only glimpsed.

Out of this marriage, Jaynes extracted a principle for determining 'how things happen' which, he argued, applied equally to equilibrium and non-equilibrium systems: entropy tends to get *maximized*. In the latter case, what this means is that a system tends to adopt the state in which entropy is produced at the *greatest* rate: that is, it does not minimize entropy production, but rather, maximizes it. This idea has not yet gained general acceptance, but there is growing support for it.

One of the attractive justifications of the theory of maximal entropy production is that it offers a rationalization of why ordered patterns may appear far from equilibrium. This, it is worth reminding ourselves once more, is a highly counter-intuitive phenomenon: we might expect systems driven out of equilibrium to dissolve into chaos. What is more, it appears to (although in fact does not) challenge the second law of thermodynamics, which insists that entropy and thus disorder must increase. And indeed, if we are now insisting that not only does entropy increase but it tends to do so at the maximum rate, why should that be a prescription for order rather than its opposite? The answer, according to Jaynes's theory of entropy maximization, is that *ordered states are more effective than disordered ones at producing entropy*. To put it another way: suppose a system has accumulated a lot of energy and 'needs' to discharge it. A rather literal expression of that situation is the build-up of electrical charge in a thundercloud, which may be released by passing electrical current to the ground. One way this could happen is for the charge to hop out onto droplets of moisture or dust in the air, and for these to gradually diffuse down to the ground. That is a slow process. What often happens instead, of course, is that the charge grounds itself all at once in a lightning bolt, creating one of the branching patterns we encountered in this book. Lightning, the dielectric breakdown of air (page 83), provides a 'structured channel' for the release of the electrical energy *at the maximal rate of entropy production*. As the physicist Roderick Dewar puts it, 'far from equilibrium, the coexistence of ordered and dissipative regions produces and exports more entropy to the environment than a purely dissipate soup'.* And so, according to Rod Swenson of the University of Connecticut 'the world can be expected to produce order whenever it gets the chance'.

The implications of this idea are extraordinary. It is one thing to explain convective roll cells and river networks as structures that arise because they offer 'channels' for relieving energy stress and producing entropy efficiently. But some researchers have gone much further than this. Harold Morowitz of George Mason University in Fairfax, Virginia, and Eric Smith of the Santa Fe Institute in New Mexico point out that life itself is an example of non-equilibrium regularity and structure, and that perhaps it is one of Swenson's inevitable ordered forms, waiting to burst forth as soon as the universe gets the chance. Morowitz and Smith argue that the early Earth was a storehouse of energy 'needing' to be dissipated. In particular, there may have been plentiful hydrogen and carbon dioxide: two molecules that release energy when they react, but which do so only very slowly on their own. Primitive living organisms would have supplied a way for this to happen, 'fixing' carbon dioxide into organic matter through reactions that use electrons extracted from hydrogen. Similarly, some geological environments generate molecules rich in electrons and others hungry for them; but only living cells would let this transfer proceed at an appreciable rate. In other words, life may have appeared on the early Earth as a kind of lightning conductor, using order to speed up entropy production. In that picture, say Morowitz and Smith, 'a state of the geosphere which includes life [was] more likely than a purely abiotic state'.

LIFE ITSELF

This is a very different view of life from the one scientists have long wrestled with. They have tended to think that, because even the most primitive organisms are hideously complicated, and because their ingredi-

---

*'Dissipative' here might be confusing in view of the discussion of 'dissipative structures' earlier. Here what it implies is a process of random rather than focused dispersal.

ents seem to be rather rare in an abiotic (inorganic) environment, life on Earth was a remarkable stroke of luck. That, however, sits uneasily with geological evidence suggesting that life probably began on our planet the instant this became geologically feasible—that is, once the surface was no longer molten, and water had condensed from the atmosphere as oceans. Moreover, the fact that life seems to thumb its nose at the second law of thermodynamics, creating order rather than succumbing to randomness, has long left scientists uneasy: the physicist Erwin Schrödinger felt forced to talk in uncomfortable terms about life as a source of 'negative entropy'. The notion of maximum entropy production and the concomitant drive towards non-equilibrium order potentially removes such paradoxes. It implies that life itself is a result of the seemingly irrepressible tendency for order to crystallize away from equilibrium.

If that is right, we have little reason to fear that we might be alone in the universe.

THE HELE-SHAW CELL

T he cell is basically two clear, rigid plates separated by a small gap. It's simplest to make these plates in the form of trays with raised edges, which keeps the liquid confined to the lower one. Glass is recommended, but clear plastic (perspex/plexiglass) works fine and is easier to use. The top tray shown here measures 27 × 27 cm, and the lower one 34 × 34 cm. The perspex is 4 mm thick, and is glued with epoxy resin.

The top plate is separated from the lower one by flat spacers at each corner—British pennies give about the right separation, as will American nickels. The viscous liquid is glycerine, bought from a pharmacist. For a readily visible and attractive pattern, you can add food colouring. (Using glycerine rather than oil makes the assembly easier to clean.) Air is injected through a small hole in the top plate; I simply drilled a hole to fit the empty ink tube from a ball-point pen (about 2 mm internal diameter). This was glued in place. It is simplest to inject the air through a

plastic syringe connected by rubber tubing; but you can just blow through the tube instead. Remember that the viscous fingering pattern is a non-equilibrium shape, which means that you need to create a substantial disequilibrium: in plain words, blow hard and sharp!

I have taken this design from:

T. Vicsek, 'Construction of a radial Hele-Shaw cell', in *Random Fluctuations and Pattern Growth*, ed. H. E. Stanley and N. Ostrowsky (Dordrecht: Kluwer Academic, 1988), p. 82.

Assenheimer, M., and Steinberg, V., 'Transition between spiral and target states in Rayleigh-Bénard convection', *Nature* 367 (1994): 345.

Audoly, B., Ries, P. M., and Roman, B., 'Cracks in thin sheets: when geometry rules the fracture path, preprint <www.lmm.jussieu.fr/platefracture/preprint_geometry_fracture.pdf>.

Avnir, D., Biham, O., Lidar D., and Malcai, O., 'Is the geometry of nature fractal?' *Science* 279 (1998): 39–40.

Ball, P., *Critical Mass* (London: Heinemann, 2004).

Barabási, A.-L., *Linked* (Cambridge, MA: Perseus, 2002).

Batty, M., and Longley, P., *Fractal Cities* (London: Academic Press, 1994).

Ben-Jacob, E., Goldenfeld, N., Langer, J. S., and Schön, G., 'Dynamics of interfacial pattern formation', *Physical Review Letters* 51 (1983): 1930.

Ben-Jacob, E., 'From snowflake formation to growth of bacterial colonies. Part I: diffusive patterning in azoic systems', *Contemporary Physics* 34 (1993): 247.

Ben-Jacob, E., 'From snowflake formation to growth of bacterial colonies. Part II: cooperative formation of complex colonial patterns', *Contemporary Physics* 38 (1997): 205.

Ben-Jacob, E., and Garik P., 'The formation of patterns in non-equilibrium growth', *Nature* 343 (1990): 523.

Ben-Jacob, E., Shochet, O., Cohen, I., Tenenbaum, A., Czirók, A., and Vicsek, T., 'Cooperative strategies in formation of complex bacterial patterns', *Fractals* 3 (1995): 849.

Ben-Jacob, E., Shochet, O., Tenenbaum, A., Cohen, I., Czirók, A., and Vicsek, T., 'Generic modelling of cooperative growth patterns in bacterial colonies', *Nature* 368 (1994): 46.

Bentley, W. A., and Humphreys, W. J., *Snow Crystals* (New York: Dover, 1962).

Bergeron V., Berger C., and Betterton, M. D., 'Controlled irradiative formation of penitentes', *Physical Review Letters* 96 (2006): 098502.

Betterton, M. D., 'Theory of structure formation in snowfields motivated by penitentes, suncups, and dirt cones', *Physical Review E* 63 (2001): 056129.

Bohn, S., Douady, S., and Couder, Y., 'Four sided domains in hierarchical space dividing patterns', *Physical Review Letters* 94 (2005): 054503.

Bohn, S., Pauchard, L., and Couder, Y., 'Hierarchical crack pattern as formed by successive domain divisions. I. Temporal and geometrical hierarchy', *Physical Review E* 71 (2005): 046214.

Bohn, S., Platkiewicz, J., Andreotti, B., Adda-Bedia, M., and Couder, Y., 'Hierarchical crack pattern as formed by successive domain divisions. II. From disordered to deterministic behavior', *Physical Review E* 71 (2005): 046215.

Bohn, S., Andreotti, B., Douady, S., Munzinger, J., and Couder, Y., 'Constitutive property of the local organization of leaf venation networks', *Physical Review E* 65 (2002): 061914.

Bowman, C., and Newell, A. C., 'Natural patterns and wavelets', *Reviews of Modern Physics* 70 (1998): 289.

Brady, R. M., and Ball, R. C., 'Fractal growth of copper electrodeposits', *Nature* 309 (1984): 225.

Brockmann, D., Hufnagel, L., and Geisel, T., 'The scaling laws of human travel', *Nature* 439 (2006): 462–465.

Brown, J. H., and West, G. B. (eds), *Scaling in Biology* (New York: Oxford University Press, 2000).

Buchanan, M., *Small World* (London: Weidenfeld & Nicolson, 2002).

Buzna, L., Peters, K., Ammoser, H., Kühnert, C., and Helbing, D., 'Efficient response to cascading disaster spreading', *Physical Review E* 75 (2007): 056107.

Chopard, B., Herrmann, H. J., and Vicsek, T., 'Structure and growth mechanism of mineral dendrites', *Nature* 353 (1991): 409.

Cohen, R., Erez, K., ben-Avraham, D., and Havlin, S., 'Resilience of the Internet to random breakdowns', *Physical Review Letters* 85 (2000): 4626–4628.

Couder, Y., Pauchard, L., Allain, C., Adda-Bedia, M., and Douady, S., *European Physical Journal B* 28 (2002): 135–138.

Cowie P., 'Cracks in the Earth's surface', *Physics World* (February 1997): 31.

Cross, M. C., and Hohenberg, P., 'Pattern formation outside of equilibrium', *Reviews of Modern Physics* 65 (1993): 851.

Czirók, A., Somfai, E., and Vicsek, T., 'Experimental evidence for self-affine roughening in a micromodel of geomorphological evolution', *Physical Review Letters* 71 (1993): 2154.

Czirók, A., Somfai, E., and Vicsek, T., 'Self-affine roughening in a model experiment in geomorphology', *Physica A* 205 (1994): 355.

Daerr, A., Lee, P., Lanuza, J., and Clément, É., 'Erosion patterns in a sediment layer', *Physical Review E* 67 (2003): 065201.

Dawkins, R., *River Out of Eden* (London: Weidenfeld & Nicolson, 1996).

Dewar, R. C., 'Maximum entropy production and non-equilibrium statistical mechanics', in Lorenz, R. D., and Kleidon, A. (eds), *Non-Equilibrium Thermodynamics and the Production of Entropy: Life, Earth, and Beyond* 41 (Berlin and Heidelberg: Springer, 2005).

Dezsö, Z., and Barabási, A.-L., 'Halting viruses in scale-free networks', *Physical Review E* 65 (2002): 055103(R).

Dimitrov, P., and Zucker, S. W., 'A constant production hypothesis guides leaf venation patterning', *Proceedings of the National Academy of Sciences USA* 103 (2006): 9363.

Dodds, P. S., Muhamad, R,, and Watts, D. J., 'An experimental study of search in global social networks', *Science* 301 (2003): 827.

Family, F., Masters, B. R., and Platt, D. E., 'Fractal pattern formation in human retinal vessels', *Physica D* 38 (1989): 98.

Fleury, V., *Arbres de Pierre* (Paris: Flammarion, 1998).

Fleury, V., 'Branched fractal patterns in non-equilibrium electrochemical deposition from oscillatory nucleation and growth', *Nature* 390 (1997): 145.

Garcia-Ruiz, J. M., Louis, E., Meakin, P., and Sander, L. M. (eds), *Growth Patterns in the Physical Sciences and Biology* (New York: Plenum Press, 1993).

Ghatak, A., and Mahadevan, L., 'Crack street: the cycloidal wake of a cylinder tearing through a thin sheet', *Physical Review Letters* 91 (2003): 215507.

Goehring, L., and Morris, S. W., 'Order and disorder in columnar joints', *Europhysics Letters* 69 (2005): 739–745.

Goehring, L., Morris, S. W., and Lin, Z., 'An experimental investigation of the scaling of columnar joints', *Physical Review E* 74 (2006): 036115.

Goehring, L., and Morris, S. W., 'Scaling of columnar joints in basalt', *Journal of Geophysical Research* 113 (2008): B10203.

Gordon, J. E., *The New Science of Strong Materials* (London: Penguin, 1991).

Gravner, J., and Griffeath, D., 'Modeling snow crystal growth II: a mesoscopic lattice map with plausible dynamics', *Physica D* 237 (2008): 385.

Gravner, J., and Griffeath, D., 'Modeling snow-crystal growth: a three-dimensional mesoscopic approach', *Physical Review E* 79 (2009): 011601.

Hurd, A. J. (ed.), *Fractals. Selected Reprints* (College Park: American Association of Physics Teachers, 1989).

Ijjazs-Vasquez, E., Bras, R. L., and Rodriguez-Iturbe, I., 'Hack's relation and optimal channel networks: the elongation of river basins as a consequence of energy minimization', *Geophysical Research Letters* 20 (1993): 1583.

Jacobs, J., *The Death and Life of Great American Cities* (New York: Vintage, 1961).

Jagla, E. A., and Rojo, A. G., 'Sequential fragmentation: the origin of columnar quasihexagonal patterns', *Physical Review E* 65 (2002): 026203.

Jeong, H., Tombor, B., Albert, R., Oltvai, Z. N., and Barabási, A.-L., 'The large-scale organization of metabolic networks', *Nature* 407 (2000): 651.

Kauffman, S., At Home in the Universe (Oxford: Oxford University Press, 1995).

Kessler, D., Koplik, J., and Levine, H., 'Pattern selection in fingered growth phenomena', *Advances in Physics* 37 (1988): 255.

Kirchner, J. W., 'Statistical inevitability of Horton's laws and the apparent randomness of stream channel networks', *Geology* 21 (1993): 591.

Landauer, R., 'Stability in the dissipative steady state', *Physics Today* 23 (November 1978).

Landauer, R., 'Inadequacy of entropy and entropy derivatives in characterizing the steady state', *Physical Review A* 12 (1975): 636.

Libbrecht, K., and Rasmussen, P., *The Snowflake: Winter's Secret Beauty* (Stillwater, MN: Voyageur Press, 2003).

Libbrecht, K., 'The enigmatic snowflake', *Physics World*, January 2008: 19.

Libbrecht, K., 'The formation of snow crystals', *American Scientist* 95(1) (2007): 52.

Liljeros, F., Edling, C. R., Nunes Amaral, L. A., Stanley, H. E., and Åberg, Y., 'The web of human sexual contacts', *Nature* 411 (2001): 907–908.

Makse, H. A., Havlin, S., and Stanley, H. E., 'Modelling urban growth patterns', *Nature* 377 (1995): 608.

Mandelbrot, B., *The Fractal Geometry of Nature* (New York: W. H. Freeman, 1984).

Mandelbrot, B., 'Fractal geometry: what is it, and what does it do?', *Proceedings of the Royal Society of London, Series A* 423 (1989): 3.

Marder, M., 'Cracks take a new turn', *Nature* 362 (1993): 295.

Marder, M., and Fineberg, J., 'How things break', *Physics Today* 24 (September 1996).

Maritan, A., Rinaldo, A., Rigon, R., Giacometti, A., and Rodriguez-Iturbe, I., 'Scaling laws for river networks', *Physical Review E* 53 (1996).

Masters, B. R., 'Fractal analysis of the vascular tree in the human retina', *Annual Reviews of Biomedical Engineering* 6 (2004): 427–452.

Matsushita, M., and Fukiwara, H., 'Fractal growth and morphological change in bacterial colony formation', in Garcia-Ruiz, J. M., Louis, E., Meakin, P., and Sander, L. M. (eds), *Growth Patterns in Physical Sciences and Biology* (New York: Plenum Press, 1993).

Meakin P., 'Simple models for colloidal aggregation, dielectric breakdown and mechanical breakdown patterns', in Stanley, H. E., and Ostrowsky, N. (eds), *Random Fluctuations and Pattern Growth* (Dordrecht: Kluwer, 1988).

Milgram, S., 'The small world problem', *Psychology Today* 2 (1967): 60.

Morowitz, H., and Smith, E., 'Energy flow and the organization of life. Santa Fe Institute Working Papers', available at <http://www.santafe.edu/research/publications/workingpapers/06-08-029.pdf>.

Müller, G., 'Starch columns: analog model for basalt columns', *Journal of Geophysical Research* 103, B7 (1998): 15239–15253.

Mullins, W. W., and Sekerka, R. F., 'Stability of a planar interface during solidification of a dilute binary alloy', *Journal of Applied Physics* 35 (1964): 444.

Mumford, L., *The Culture of Cities* (London: Secker and Warburg, 1938).

Murray, A. B., and Paola, C., 'A cellular model of braided rivers', *Nature* 371 (1994): 54.

Nicolis, G. 'Physics of far-from-equilibrium systems and self-organization', in Davies, P. (ed.), *The New Physics* (Cambridge: Cambridge University Press, 1989).

Niemeyer, L., Pietronero, L., and Wiesmann, H. J., 'Fractal dimensions of dielectric breakdown', *Physical Review Letters* 52 (1984): 1033.

Nittmann, J., and Stanley, H. E., 'Tip splitting without interfacial tension and dendritic growth patterns arising from molecular anisotropy', *Nature* 321 (1986): 663.

Nittmann, J., and Stanley, H. E., 'Non-deterministic approach to anisotropic growth patterns with continuously tunable morphology: the fractal properties of some real snowflakes', *Journal of Physics A* 20 (1987): L1185.

Oikonomou, P., and Cluzel, P., 'Effects of topology on network evolution', *Nature Physics* 2 (2006): 532.

Pastor-Satorras, R., and Vespignani, A., 'Epidemic spreading in scale-free networks', *Physical Review Letters* 86 (2001): 3200.

Pastor-Satorras, R., and Vespignani, A., 'Optimal immunisation of complex networks', preprint <http://www.arxiv.org/abs/cond-mat/0107066 (2001)>.

Perrin, B., and Tabeling, P., 'Les dendrites', *La Recherche* 656 (May 1991).

Prigogine, I., *From Being to Becoming* (San Francisco: W. H. Freeman, 1980).

Prusinkiewicz, P., and Lindenmayer, A., *The Algorithmic Beauty of Plants* (New York: Springer, 1990).

Rigon, R., Rinaldo, A., and Rodriguez-Iturbe, I., 'On landscape self-organization', *Journal of Geophysical Research* 99 (B6) (1994): 11971.

Rinaldo, A., Banavar, J. R., and Maritan, A., 'Trees, networks, and hydrology', *Water Resources Research* 42 (2006): W06D07.

Rodriguez-Iturbe, I., and Rinaldo, A., *Fractal River Basins. Chance and Self-Organization* (Cambridge: Cambridge University Press, 1997).

Rodriguez-Iturbe, I., Rinaldo, A., Rigon, R., Bras, R. L., Ijjasz-Vasquez, E., and Marani, A., 'Fractal structures as least energy patterns: the case of river networks', *Geophysical Research Letters* 19 (1992): 889.

Ryan, M. P., and Sammis, C. G., 'Cyclic fracture mechanisms in cooling basalt', *Geological Society of America Bulletin* 89 (1978): 1295.

Sapoval, B., Baldassarri, A., and Gabrielli, A., 'Self-stabilized fractality of seacoasts through damped erosion', *Physical Review Letters* 93 (2004): 098501.

Sapoval, B., *Universalités et Fractales* (Paris: Flammarion, 1997).

Sander, L. M., 'Fractal growth', *Scientific American* 256(1) (1987): 94.

Shorling, K. A., Bruyn, J. R. de, Graham, M., and Morris, S. W., 'Development and geometry of isotropic and directional shrinkage crack patterns', *Physical Review E* 61, 6950 (2000).

Skjeltorp, A., 'Fracture experiments on monolayers of microspheres', in Stanley, H. E., and Ostrowsky, N. (eds), *Random Fluctuations and Pattern Growth* (Dordrecht: Kluwer, 1988).

Sinclair, K., and Ball, R. C., 'Mechanism for global optimization of river networks from local erosion rules', *Physical Review Letters* 76 (1996): 3360.

Stanley, H. E., and Ostrowsky, N. (eds), *On Growth and Form* (Dordrecht: Martinus Nijhoff, 1986).

Stanley, H. E., and Ostrowsky, N. (eds), *Random Fluctuations and Pattern Growth* (Dordrecht: Kluwer, 1988).

Stark, C., 'An invasion percolation model of drainage network evolution', *Nature* 352 (1991): 423.

Stewart, I., *What Shape is a Snowflake? Magical Numbers in Nature* (London: Weidenfeld & Nicolson, 2001).

Swenson, R., 'Autocatalysis, evolution, and the law of maximum entropy production: a principled foundation towards the study of human ecology', *Advances in Human Ecology* 6 (1997): 1–47.

Swinney, H., 'Emergence and the evolution of patterns', in Fitch, V. L., Marlow, D. R., and Dementi, M. A. E. (eds), *Critical Problems in Physics* (Princeton: Princeton University Press, 1997).

Temple, R. K. G., *The Genius of China* (London: Prion Books, 1998).

Thompson, D'A. W., *On Growth and Form* (New York: Dover, 1992).

Van Damme, H., and Lemaire, E., 'From flow to fracture and fragmentation in colloidal media', in Charmet, J. C., Roux, S., and Guyon, E. (eds), *Disorder and Fracture* (New York: Plenum Press, 1990).

Vella, D., and Wettlaufer, J. S., 'Finger rafting: a generic instability of floating ice sheets', *Physical Review Letters* 98 (2007): 088303.

Watts, D. J., and Strogatz, S. H., 'Collective dynamics of "small-world" networks', *Nature* 393 (1998): 440–442.

Watts, D. J., *Small Worlds* (Princeton: Princeton University Press, 1999).

Watts, D. J., Dodds, P. S., and Newman, M., 'Identity and search in local networks', *Science* 296 (2002): 1302–1305.

Watts, D. J., *Six Degrees* (New York: W. W. Norton, 2004).

Weaire, D., and O'Carroll, C., 'A new model for the Giant's Causeway', *Nature* 302 (1983): 240–241.

West, G. B., Brown, J. H., and Enquist, B. J., 'A general model for the origin of allometric scaling laws in biology', *Science* 276 (1997): 122.

Whitfield, J., *In the Beat of a Heart* (Washington: Joseph Henry Press, 2006).

Yuse, A., and Sano, M., 'Transitions between crack patterns in quenched glass plates', *Nature* 362 (1993): 329.